Modeling Reality with Functions: Graphical, Numerical, Analytical

Gene Fiorini, Ph.D.
Department of Mathematics & Computer Science
Shippensburg University
Shippensburg, PA

Jacob Miller, Ph.D.
Department of Information Sciences
Pennsylvania College of Technology
Williamsport, PA

KENDALL/HUNT PUBLISHING COMPANY
4050 Westmark Drive   Dubuque, Iowa 52002

iii

Abstract

This text is uses an application-based approach to general education-level mathematics. It is designed to develop quantitative thinking skills in students at the freshman/sophomore level of mathematics. The concepts are presented through a series of problems that are based on actual data and situations faced by businesses, industry, government and media sources. These problems demonstrate the relationships between geometric, analytical and numerical models as well as how to develop these mathematical models.

The text is also designed to be interactive. Students and instructors are encouraged to write answers to section exercises in the spaces and grids provided. Although many problems and examples are contained in each section instructors should feel free to diverge from prepared exercises in favor of their own related exercises and problems that they have developed from their own sources.

Contents

0.1	Preface	i

1 Models and Functions **1**
1.1 Introduction: What are Mathematical Models? 1
1.2 Geometric, Analytical, and Numerical Models 2
    1.2.1 An Example of a Geometric Model: Tracking the Spread
         of a Disease 3
    1.2.2 An Example of an Analytical Model: A Buried Fuel Tank 9
    1.2.3 An Example of a Numerical Model: Handprints 11
    1.2.4 Problems 14
1.3 Role of a Function in Models 20
    1.3.1 Definition of a Function 20
    1.3.2 Notation: A Purely Numerical Example 23
    1.3.3 Measuring the Severity of a Hurricane: An Application . 25
    1.3.4 Problems 28
1.4 Summary 33

2 Linear Function Models **35**
2.1 Introduction 35
2.2 Rates of Change 36
    2.2.1 Speed 36
    2.2.2 Problems 41
2.3 Linear Rates of Change 43
    2.3.1 Geometric and Analytic Representations of Linear Models 43
    2.3.2 Numerical and Analytic Representations of Linear Models 48
    2.3.3 Problems 53
2.4 The Power of Linear Models: More Applications 62
    2.4.1 Step Functions Models 62
    2.4.2 Systems of Linear Models 66
    2.4.3 Models of Linear Inequalities 69
    2.4.4 Linear Programming Models 72
    2.4.5 Problems 76
2.5 Summary 78

3 Models of Ratio and Proportion   **79**
   3.1 Introduction 79
   3.2 Population Characteristics 80
       3.2.1 Tagging: Estimating Population Size 80
       3.2.2 Analyzing Population Behavior 83
       3.2.3 Problems 86
   3.3 Direct Comparisons 90
       3.3.1 Measuring Diversity 90
       3.3.2 Measuring Porosity 93

       3.3.3 Measuring Height 96
       3.3.4 Problems 103
   3.4 Inverse Proportions 108
       3.4.1 Highway Safety 108
       3.4.2 Problems 112
   3.5 Summary 114

4 Quadratic Function Models   **115**
   4.1 Introduction 115
   4.2 Finding a Quadratic Model to Fit Data 116
       4.2.1 Finite Differences Applied to Quadratic Data 116
       4.2.2 Finite Differences Applied to Quadratic Graphs 117
       4.2.3 Finite Differences Applied to Quadratic Regression 120
       4.2.4 Problems 123
   4.3 Multiplying Binomials (FOIL) 130
       4.3.1 The Tree and Rectangular Methods of Multiplying Poly-
              nomials 131
       4.3.2 Applying the Tree Method 133
       4.3.3 Problems 135
   4.4 Factoring Quadratic Models 136
       4.4.1 Factoring Techniques. 136
       4.4.2 Applications of Factoring to Models 141
       4.4.3 The Quadratic Formula and Completing the Square . . . 144
       4.4.4 Problems 147
   4.5 Translation of Axes 149
       4.5.1 Problems 152
   4.6 Summary 154

5 Models Using Averages   **155**
   5.1 Introduction 155
   5.2 The Coordinate Plane 155
   5.3 The Arithmetic Average 157
       5.3.1 One-Dimensional Averages 157
       5.3.2 Two-Dimensional Averages 159
       5.3.3 Problems 163
   5.4 The Geometric and Harmonic Averages 166

5.4.1	The Geometric Average	. . . .	166
5.4.2	The Harmonic Average	. . . .	166
5.4.3	Problems	. . . .	167
5.4.4	Linear Regression	. . . .	168
5.4.5	Problems	. . . .	171
5.5	Summary	. . . .	173

6 Models Using Polynomials   **175**

6.1	Introduction	. . . .	175
6.2	Simplifying Expressions	. . . .	175
	6.2.1 Properties of Exponents	. . . .	176
	6.2.2 Combining Like Terms	. . . .	178
	6.2.3 Problems	. . . .	181
6.3	Polynomial Models	. . . .	183
	6.3.1 A Model Involving Dice	. . . .	183
	6.3.2 Polynomial Models Applied to Decision Making	. . . . .	188
	6.3.3 Polynomial Models Applied to Storage	. . . .	190
	6.3.4 Problems	. . . .	193
6.4	Polynomials and Codes	. . . .	195
	6.4.1 An Example of Modulo Arithmetic	. . . .	195
	6.4.2 Modulo Arithmetic Operations	. . . .	195
	6.4.3 Applications	. . . .	198
6.5	Summary	. . . .	201

7 Exponential Models   **203**

7.1	Introduction	. . . .	203
7.2	Financial Investment Models	. . . .	205
	7.2.1 Start Investing Early: Compound Interest	. . . .	205
	7.2.2 Start Investing Early: Periodic Deposits	. . . .	207
	7.2.3 Carbon Dating	. . . .	211
	7.2.4 Problems	. . . .	213
7.3	Summary	. . . .	216

Additional Grids   **217**

Additional Charts   **221**

0.1 Preface

When students in a traditional algebra class are assigned word problems, if they are assigned word problems, a standard type of problem they are asked to solve is the following.

Find two consecutive positive integers whose product is 5402.

We are probably all familiar with the standard algorithmic approach used by most instructors:

1. Students are instructed to let the variable x represent the smaller of the two positive integers. Then $x + 1$ represents the larger of the two integers.

2. Students are then told to multiply x and $x + 1$ and set the product equal to 5402.

$$x(x + 1) = 5402 \tag{1}$$

3. Multiplying the left side of 1 students obtain

$$x^2 + x = 5402 \tag{2}$$

4. By subtracting 5402 from both sides of 2 the students quickly arrive at the quadratic equation

$$x^2 + x - 5402 = 0.$$

5. To solve this quadratic students are instructed that to factor this quadratic they should employ a type of reverse FOIL. That is, find those factors of the constant term whose factors differ by the linear coefficient. In this case,

find those factors of 5402 that differ by the quantity one!

At the conclusion of such a futile exercise, students may justifiably ask "*Why do we need to know this?*" This is a song that is sung many times over in mathematics classrooms throughout this country. The usefulness and meaning of mathematics has been obscured behind an esoteric algorithmic process that leaves students wondering what purpose mathematics serves. In an age in which technology is taking a more prominent role in our every-day lives we can no longer afford to alienate students from the mathematical process through meaningless, algorithmic procedures.

The purpose of this text is to introduce some of the basic mathematical concepts through an applications-based approach. That is, the concepts outlined in the text are introduced through problems that have been derived from actual situations. Since its publication in 1989 the National Council of Teachers of

Mathematics *Principles in Teaching and Curriculum Standards* has advocated incorporating applications into the mathematics classroom as one method in the efforts to reform mathematics teaching and curriculum. However, most of the efforts to incorporate applications into mathematics lessons has taken one of two forms. Either the applications are introduced only after the concepts have been thoroughly reviewed or applications are used in the *discovery* method where students are asked to "discover" the concept through an in-class activity that leads them to the inevitable conclusion. In the case of the former approach, the applications are given short attention or eliminated altogether if the instructor is *"running out of time"* and the syllabus items need to be covered before the term ends. In the latter approach it is often argued that the discovery method is most effective at the beginning stages of mathematics. Thus, discovery activities are usually limited to basic, simple concepts that require only a few simple steps or noadvanced mathematical ideas.

Both approaches may also be hampered by an instructor's belief that "real" applications require higher-level mathematics or can only be competently solved by the *better* students. Thus, applications-based activities are postponed until students have "sufficient" mathematical background. In their place are contrived, simpler examples that deal with the current topic. In fact, there are many examples of actual problems from industry, business and science as well as other sources that use some of the most basic mathematical concepts. This text outlines many of these problems and uses them as a means to introduce the topics to the students as well as give the concepts relevance. Although some of the problems that appear in the text have been slightly altered for effect, most of the problems were developed from actual sources in industry, science or business. As a result solutions to problems may be somewhat ambiguous. This is deliberate so as to encourage a dialogue between the students as well as between the instructor and the students in the hope that these dialogues will lead to a clearer understanding of the concepts being used to solve the problem. This also reflects the "real-world" situation that there may be more than one correct approach to solving a problem.

It is recommended that Chapters One and Two be covered in their entirety. Chapter One introduces the concept of mathematical models. Mathematical modeling is the use of mathematics to represent or simulate an actual situation by graphical, numerical or analytical means or some combination thereof. Both the advantages and disadvantages to graphical, numerical and analytical models are discussed as well as how combinations of these three representations are used to give a clearer picture of the situation. Chapter Two gives a detailed discussion of linear models from graphical, numerical and analytical points of view. As one of the simplest and most useful mathematical models students should have clear understanding of how to apply linear models.

As for the rest of the text, the remaining chapters are designed to be covered independently of each other. Instructors need not cover the remaining chapters in succession and are encouraged to "skip around" creating a course that will be of maximum benefit to their students. Since every class of students is unique, instructors are also encouraged to alter given examples or create their own activities so that they are more relevant to their students' backgrounds and needs. This approach may require additional preparation time in the beginning, however, the rewards from such an approach will more than compensate any additional time spent on the developments of activities. If the instructor has had little experience developing their own activities the authors offer some helpful hints here to getting started.

- Initiate an on-going dialogue with colleagues, both mathematical colleagues as well as colleagues in other fields, as to how to approach a particular topic or concept.

- Such a dialogue among colleagues works best in an environment that encourages an exchange of ideas. Find a suitable venue that will allow for an exchange of ideas and make these meetings and discussions a part of your regular schedule.

- Develop contacts beyond the academic environment (industry, business, media, etc.) that are willing to have an on-going dialogue and an exchange of ideas that will serve as a resource for new and current ideas.

- Do not hesitate to abandon "old standards" if they are not working with the current class of students. Each new class of students brings with them their own experiences which changes from class to class. What is relevant to one class may not be relevant to another class.

- Do not hesitate to revisit ideas. An idea that does not work with one class may work just fine with another. Also, an idea that goes nowhere with one group of colleagues may be just the thing with another group of colleagues.

- Search existing sources of material for ideas, however, do not hesitate to alter an existing activity to fit the needs of your current class.

Since the authors wish to encourage a free exchange of ideas, no solution guide is included with the text. However, the authors are also aware of the fact that many instructors would like to have some guidelines as to what would be "correct" approaches to a problem. As a result a solutions manual is available on the internet. Although this manual does not include specific numerical answers to many of the problems, it does include guidelines on various approaches to problems as well as supporting documentation and additional resources. The internet site is located at *http://www.ship.edu/~grfior/texthelp.html* and includes a "chat room" for instructors to exchange ideas as well as offer each other assistance.

Special thanks go to Dr. Andre Acusta of SmithKline Beecham Pharmaceutical Research and Development and Dr. Sandra Gorka of the Pennsylvania College of Technology for all their help, support and suggestions. The advice and suggestions made by Dr. Acusta and Dr. Gorka were invaluable. Many of the applications and problems that appear in this text were developed from their suggestions and experience. Their willingness to sit through endless discussions over cold dinners deserves special consideration. We are also indebted to Dr. Jack Mowbray of Shippensburg University. He willingly volunteered his classes as an environment to test our ideas. This has earned him our admiration and gratitude. We owe a great deal to his generosity and patience. His suggestions were insightful and have improved the quality of the text.

Chapter 1

Models and Functions

1.1 Introduction: What are Mathematical Models?

In general, a **model** is a small object, usually built to scale, that represents another, often larger object. Of course, by "small" we mean in relation to the size of the actual object it is to represent. The reasons for constructing models are many. Certain toys can be thought of as models: dolls and their accessories, toy cars, toy planes, etc. Such toys serve not only for amusing children but also serve as a means of child instruction and development. Architects and artists often build scale models first to test structural integrity of building and sculpture designs. Automobile manufacturers test the aerodynamics of new designs before the plans are sent into production. These are but a few of the ways in which models are used as representations of actual objects. Clearly, models are used as an efficient way of representing certain aspects of the real world that we wish to study, alter or improve.

Models, however, need not be concrete objects in order for us to study the world. There are also many examples of abstract models, preliminary patterns that serve as plans from which as yet unconstructed items will be produced. A building's blueprints are a simple example of this. It is within this category of abstract models that the concept of mathematical modeling falls.

Definition 1 A *mathematical model* is a combination of mathematically related functions, graphs, equations, charts, etc. that represent specific characteristics of a real situation or world problem that is to be studied and solved.

Of course, no model, concrete or abstract, duplicates the actual object or situation exactly in every detail. Due to constraints on the object's or problem's variables many of the original item's characteristics are lost or abandoned in the model. This is only practical. Seldom will all the characteristics be needed in order to successfully develop a workable solution. We are generally not interested in studying all the characteristics of an item at once. In order to successfully

1

create a mathematical model, the modeler must, first learn how to identify and isolate those characteristics they wish to study, and then learn how to construct a working model that contains those characteristics.

In this chapter we will introduce the reader to some of the basic procedures of modeling. This text is designed to assist the beginning modeler in identifying important characteristics of real-world problems as well as building concise and efficient mathematical models to solve those problems. We will look at several basic mathematical model types and their uses. All the concepts are introduced through the use of actual problems that were gathered from local industry (businesses, banking, media, government agencies, etc.) sources in the region.

1.2   Geometric, Analytical, and Numerical Models

There are three primary mathematical model representations of problems: geometric, analytical, and numerical. There are advantages and disadvantages to using each of these representations. That is why some combination of these three models is used when representing more complicated problems or when a more accurate representation is required. Combining together different model types to represent problems takes practice and experience. Before trying to combine model types together we will demonstrate each model type by way of simple examples.

Remark 2 *There are several suggestions that can be used to help students when creating a model.*

1. *Identify which questions are to be answered by the model. The fewer questions the better. Since more characteristics required more complicated models it is better to limit an experiment to answer as few questions as possible. In fact, it is best to try to answer just **one** question. If mathematical consultants are interested in answering several questions, then different models can be developed for each of those questions.*

2. *Once the question(s) to be answered are clearly stated, then the type(s) of models that will best answer these questions can be identified. Determine which model, through its advantages and disadvantages, can best represent the question(s) to be answered.*

3. *Use your knowledge of mathematics and the available information to represent the problem by the appropriate model.*

4. *Use this information to solve any equations, graphs, functions, etc. in order to obtain the desired answer.*

As we progress through this text we will be more specific as to how to how to proceed through each of the steps above. For now we will simply demonstrate some of the general techniques through some simple examples.

1.2.1 An Example of a Geometric Model: Tracking the Spread of a Disease

When a problem is represented using a graph, chart, diagram or some other geometric figure, the figure is referred to as a **geometric model** of that problem.

Remark 3 *There are several advantages to using geometric models.*

- *Visual representations of problems communicate major characteristics quickly, especially to a large (or non-expert) audience.*

- *The geometric representation gives a general, overall view of those characteristics: trends, highs and lows, critical points, etc.*

- *General estimates of values can be obtained quickly through visual inspection.*

Use the space below to write down some additional advantages to using geometric models.

- Additional Advantages:

Remark 4 *There are several disadvantages to using geometric models as well.*

- *Geometric representations of problems are inaccurate. They can only give an approximation of critical values.*

- *Greater precision cannot be obtained beyond the axes' scale.*

- *Original data cannot always be retrieved from the geometric model.*

Use the space below to write down some additional disadvantages to using geometric models.

- Additional Disadvantages:

As an example of a geometric model consider the Centers for Disease Control (CDC) located in Atlanta. Among other things the CDC's job is to monitor the spread of known diseases that could pose a public health hazard, identify and catalog the rise of new diseases, as well as educate the public of any new threat to the safety of their health. It is not enough, however, to simply wait for a disease to occur in the general public in order for the CDC to gather necessary data on it. To increase public safety the CDC must anticipate the occurrence of diseases. Specific symptoms of diseases can be recreated in a laboratory environment under carefully controlled conditions. Within this safe environment scientists and medical experts can determine answers to such questions as how a disease spreads from one host to another, how fast the disease spreads, how deadly it is, etc.

A well-equipped laboratory can be very helpful to answering many of these questions but is not necessary for some of the questions. The beauty of mathematics is that it can be done anywhere. Its laboratory is in the mind. As mathematical modelers we can simulate certain characteristics of diseases by purely mathematical methods. For example we can simulate the speed at which a disease will spread by making certain assumptions about how people "catch" the disease.

Experiment: Fifty people work within a closed environment in a building. Throughout the day people come in contact with each other. Let's suppose that one of the fifty people comes to work that day with the common cold.

Exercise 5 *This simple scenario presents the medical consultant with many possible questions. By placing yourself in the position of a consultant use the space below to list some possible questions that you would want answered about the cold experiment above.*

For the purpose of our example we will be interested in answering the question: How long it will take for the cold to spread through the fifty people? Of course, along the way we may also be able to answer some other questions about the spread of the cold. In reality just because we come in contact with someone with a cold does not mean we will catch that cold, or if we do catch it that the symptoms will be exactly the same. In order to better represent the experiment with a mathematical model we will make some assumptions that will simplify our computations. Simplifying assumptions are often necessary in order that mathematical computations be possible. This also means that our final solution will not be a "true" solution but only an approximation, however, the approximation should still be "good enough" to answer our question(s). So, for the purposes of this experiment we will make the following assumptions:

- Everyone's resistance is low and that simply coming in contact with an infected person is enough to catch cold.

- Each person in the building will come in contact with another person in the building every half hour.

- We will assume a "closed" environment. That is, throughout the eight hour work day no one will enter or exit the building and any "outside" influences will be ignored.

- Once a person has been infected they can then pass the cold along to the person they meet during the next half hour interval. That is, the incubation period time is zero.

This is enough information to generate a graphical model. We can simulate the spreading of this cold by using a random number generator. Random number generators are found on most calculators with advanced function abilities. If such a calculator is not available then more traditional methods can be used such as a random number table or by simply pulling numbers out of a hat. Here's how we will do it.

Exercise 6 *To generate data showing how far the cold has spread over each half hour we will generate the same number of random numbers as the number of people infected as of the previous half hour. That is,*

1. *Identify each of the people by numbering them from 1 to 50. The original person who came to work with a cold will be numbered 1. To help you in this exercise the numbers from 2 to 50 are listed below. To keep track of the numbers that are generated and those numbers that remain refer to this list and cross off each newly randomly generated number.*

2. *Each new infected case will be identified by a new randomly generated number. Since one person was infected at the start of the day, during the first half hour they will come in contact with one additional person and infect that person with their cold. So, for the first half hour randomly generate one number from 1 to 50. Record your results in the chart below.*

3. *Since two people are now infected after the first half hour, each will come in contact with an additional person and infect them with their cold. So, for the second half hour generate two numbers from 1 to 50 randomly.*

If a number occurs that was previously generated then do not record this number again. Once a person is infected they cannot be infected again. Only record the newly infected numbers in the chart below.

4. *Repeat this process for each half hour in the eight hour day (a total of 16 times). For each half hour generate as many numbers from 1 to 50 randomly as there were (total) infected cases as of the previous half hour. Record the results below.*

Time	List of New Cases	# New Cases	Total Infected
0.0	1	0	1
0.5		1	2
1.0			
1.5			
2.0			
2.5			
3.0			
3.5			
4.0			
4.5			
5.0			
5.5			
6.0			
6.5			
7.0			
7.5			
8.0			

```
 1   2   3   4   5   6   7   8   9  10
11  12  13  14  15  16  17  18  19  20
21  22  23  24  25  26  27  28  29  30
31  32  33  34  35  36  37  38  39  40
41  42  43  44  45  46  47  48  49  50
```

We will now use this data to create our geometric model.

Exercise 7 *On the grid below plot the data (time vs. total infected) that was generated above.*

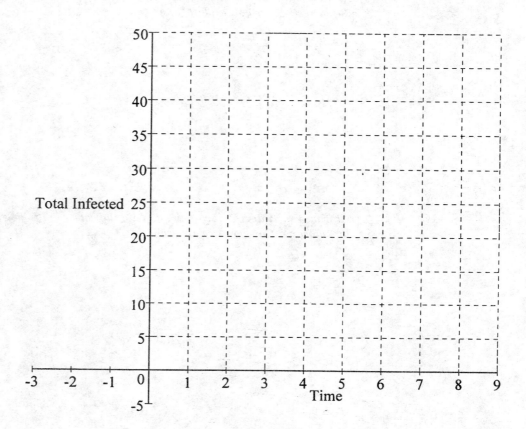

Total Infected vs. Time

Exercise 8 *Spend time in class answering each of the following questions using the data generated above.*

1. *Sketch a curve in the grid above that best follows the path of the dots. (In future chapters we will discuss how to analytically find an expression of a curve that best fits the plot data. The study of finding a curve that best fits the data is called regression.)*

2. *Spend some time in class to best answer each of the following questions (refer to the information in the above exercises as needed).*

   (a) *How long did it take for the cold to spread to most of the people in the building?*

(b) At what time interval during the day did the cold seem to be spreading the fastest? Can you explain why?

(c) At what time interval during the day did the cold seem to be spreading the slowest? Can you explain why?

(d) We mentioned earlier that no model will contain every characteristic of a situation. Thus, there are some questions that cannot be answered by each model. Make a list of questions below that, as a medical expert, you are interested in answering about this cold but that the above (graphical) model cannot answer.

1.2.2 An Example of an Analytical Model: A Buried Fuel Tank

When a problem's solution is obtained through a standard formula or equation the process is referred to as an **analytical model** of that problem.

Remark 9 *There are several advantages to using analytical models.*

- *Formulas and equations provide exact solutions to problems.*

- *Once the formulas and equations are identified there are well-known, standard techniques that can be employed to solve the problem.*

Use the space below to write down some additional advantages to using analytical models.

- Additional Advantages:

Remark 10 *There are several disadvantages to using analytical models as well.*

- *Determining the correct formulas and equations can be difficult.*

- *Analytical models are information-intensive. That is, a great deal of information is required to set up the appropriate equations and formulas. This information may not always be available.*

Use the space below to write down some additional advantages to using geometric models.

- Additional Disadvantages:

As an example of a analytic model consider a common problem that occurs when construction crews must demolish a building. Often fuel tanks, storage tanks, pipes, etc. buried beneath buildings during the construction of the building. This can present a problem when it is decided to demolish the building. Storage tanks can be a source of destabilization endangering the lives of the construction crew during demolition. The following example is a common scenario.

Experiment You are in charge of a cleanup crew which is excavating a collapsed building. Your crew uncovers the end of a cylindrical natural gas tank in the wreckage (see diagram 1.1). If the tank is longer than about 10 meters, removing it will destabilize part of the rubble causing further collapse. The end of the tank is about 2 meters in diameter. A stamping on the end of the tank indicates it has a 30 cubic meter capacity.

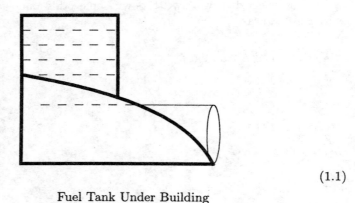

(1.1)

Fuel Tank Under Building

1. Use the space below to calculate the length of the tank. Explain which formula(s) are being used and why.

2. Is your answer for the length of the tank an exact answer or an approximation? Support your answer with an explanation.

3. Explain if you think it is possible to safely remove the tank.

1.2.3 An Example of a Numerical Model: Handprints

When a problem's solution is obtained through algorithmic processes, that is the use of repetitive computations, the process is referred to as a **numerical model** of that problem.

Remark 11 *There are several advantages to using numerical models.*

- *Numerical modeling is often not information-intensive. That is, very little information about the situation is needed in order to derive a solution.*

- *Numerical modeling can usually be expanded to include a supporting graphical model.*

Use the space below to write down some additional advantages to using numerical models.

- Additional Advantages:

Remark 12 *There are several disadvantages to using numerical models as well.*

- *Numerical modeling is inaccurate though approximations are more accurate that geometric modeling.*

- *Numerical modeling can be time consuming due to repetitive computations.*

Use the space below to write down some additional advantages to using numerical models.

- Additional Disadvantages:

Not all our examples need be serious applications to industry. For an example of a numerical model we will have some fun.

Experiment Place as much of your right hand as possible on the blank page that appears after this experiment. Using your pencil or pen trace the outline of your hand on the page. When you have completed the trace work in groups and answer each of the following questions.

1. Obviously, the shape of your hand outline is not a simple geometric shape. Thus there is no simple analytical formula that can be used to find its area. Discuss among your group possible ways to estimate the area of your hand outlines. Describe your method here.

2. Once this is done compute the hand outline area.

3. Explain how accurate your answer is.

4. Discuss ways that you might be able to improve on the accuracy of your answer.

Hand Outline Page

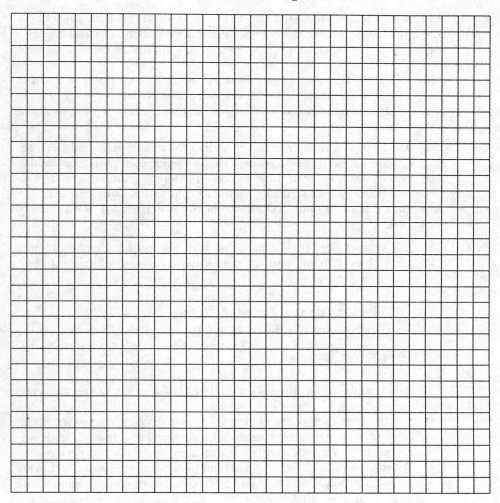

1.2.4 Problems

1. In section 1.2.1 we simulated the spread of a cold through an office building. In doing this we made some assumptions about the resistance of people in the building to a cold as well as how often they came in contact.

   (a) Explain if the assumptions we made in 1.2.1 about how the cold will spread are realistic or not. If not, what sort of changes can you make in the assumptions so that the model will reflect a more realistic situation.

   (b) Most people's resistance is higher than the assumption made in 1.2.1. To correct for this let's assume that a noninfected person must have two contacts with an infected person before catching the cold. This can be simulated again by randomly generating numbers, only this time a number must occur twice before it is recorded as an infected case.

      i. Repeat the "cold" simulation from section 1.2.1 using this new assumption and record the information in a table similar to table 4.

      ii. Plot the data (total infected vs. time).

      iii. Draw some conclusions about the speed at which the cold is spreading in this case. That is, when is the cold spreading the fastest? slowest? etc.

2. Use the example of the spread of a cold in 1.2.1 for each of the following:

   (a) Create your own scenario about how high people's resistance is or how often they come in contact with each other. (Hint: To make the model more realistic construct a scenario in which the people have different resistance factors. This can be done, for example, by having people 1 through 25 becoming infected after the initial contact and people 26 through 50 becoming infected after the second contact. Another possibility would be to allow people to become re-infected after a period of time. Discuss in class ways in which this could be accomplished.)

   (b) Use your assumptions in part a. and answer each of the following questions.

      i. Randomly generate data using these new assumptions and record the information.

      ii. Plot the data (total infected vs. time).

      iii. Draw some conclusions about the speed at which the cold is spreading in this case. That is, when is the cold spreading the fastest? slowest? etc.

(c) Scientists in the labs of the CDC cannot afford to simply guess at the resistance factor of the public. Offer some suggestions as to how the CDC could actually determine, with some accuracy, just how high people's resistance to diseases would be.

3. When a product, such as an automobile or computer model, is removed from production there continues to be a demand for replacement parts and spare parts since people continue to use the products already in the marketplace. Therefore companies, such as Pep Boys and NAPA Auto Parts in the case of cars, continue to supply spare parts for these products well after production has stopped. As time passes and fewer of the terminated products remain in use there is less need for such spare parts. There will always be some need for spare parts, however, since collectors and restorers will be interested in maintaining classic products. The following graph 1.2 models the supply of spare parts over a period of years. Use the graph model 1.2 to answer the following questions

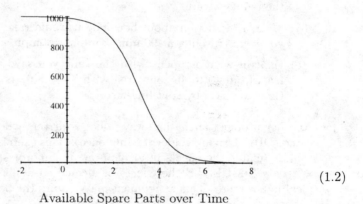

(1.2)

Available Spare Parts over Time

(a) Approximately how many spare parts are available when production of the product first ceases? Explain.

(b) Approximately how many spare parts are available 5 years after production ceases? Explain.

(c) Over what time interval is availability of spare parts decreasing the fastest? Explain.

(d) What does this model NOT tell you about the availability of spare parts?

4. Road rage has been a recurring item on newscasts over the last few years. Sometimes drivers take out their frustrations of the day by driving aggressively. Many people think that time is saved by driving faster. Although this is true, even if we ignore the safety factors, the time that is saved is relative to the speed at which the car is traveling. We will test this theory

in this problem. Suppose that we want to determine how long it will take
to cover 20 miles on an interstate highway traveling at various speeds.

(a) Write down the formula that relates distance, speed (rate) and time.
For this exercise we will assume that $D = 20miles$. Use this formula
to fill in the chart below with the times that it will take to travel 20
miles that correspond to the give speeds.

RATE (SPEED)	40	45	50	55	60	65	70	75	80	85
TIME										

You can rely on a calculator to perform all necessary calculations.

(b) Plot the data on a grid. Label the vertical axis as TIME and the
horizontal axis as RATE.

(c) Examine the pattern of points closely. Sketch in a curve that models
the point pattern.

(d) Make a prediction about how long it would take to travel 20 miles if
you were traveling at 90mph. 35mph. 100mph. Explain.

(e) Explain what happens with the time you save as you travel faster
(say, from 40 to 45 compared with 75 to 80). Is it always worth the
time you save by traveling faster?

5. Another problem facing highway police is drivers who drive while intoxi-
cated (DUI - Driving Under the Influence). This exercise will build a model
that simulates the effects of alcohol on your motor skills. Of course, we
will only SIMULATE the effects of alcohol so don't make a run to the
local liquor store. This assignment may require the help of a friend. Here
is a way to simulate the effects of alcohol.

(a) Setting up the experiment:

    i. Acquire a box of ordinary paper clips. Place them in piles of
about 15 paper clips each.

    ii. Using the paper clips from the first pile, see how many of the
paper clips you can string together in 15 seconds. (Have a friend
time you on a stop watch.)

    iii. Using the second pile of paper clips, spin around in place as fast
as you can **five** times then see how many of the paper clips you
can string together in 15 seconds. (Again have a friend time you
on a stop watch.) Record the results.

    iv. Using the third pile of paper clips, spin around in place as fast
as you can **ten** times then see how many of the paper clips you
can string together in 15 seconds. Record the results.

v. Continue this process for as long as you can. Using the nth pile of paper clips, spin around in place as fast as you can **5n** times then see how many of the paper clips you can string together in 15 seconds. Record the results. If you have to stop due to excessive dizziness, feeling sick, unable to sting any paper clips together, then record 0 as your final result.

- *Note to the instructor: This problem makes a good in-class exercise. Students can compare results.*

(b) Constructing the model:

i. Place your results in the table below.

Number of Spins	0	5	10	15	20	25	30	35	40	45
Length of String										

ii. Plot the data on a grid. Label the vertical axis as LENGTH OF PAPER CLIP STRING and the horizontal axis as NUMBER OF SPINS.

iii. Examine the pattern of points closely. Sketch in a curve that models the point pattern.

iv. Interpret your results as to how many "drinks" (5 spins = one drink) it takes to impair your motor skills. Explain.

v. Every model has built-in errors. This is due to approximations that are made, bad data, uncontrollable or unpredictable events in the experiment. Explain some of the ways your model could be in error. Include some questions about drinking that cannot be answered by this model.

6. On your way from class stop and pick up an average sized leaf. The surface area of a leaf is essential in determining the amount of sunlight it can absorb.

(a) Place the leaf on a blank sheet of paper and trace its outline.

(b) Use the method you outlined in section 1.2.3 to estimate the area of the leaf.

(c) Explain how you might be able to improve on the accuracy of your answer.

7. City planners have decided to place a new sidewalk around the city's park. The park is one square city block in the downtown area. The square has been measured to be about 21 meters in length. In order to provide support for the sidewalk and proper rainwater drainage it has been determined that there should be 10 centimeters of gravel underneath a sidewalk that will be 5 centimeters in depth.

(a) If the sidewalk is to be 1 meter in width, determine the amount of cement that has to be ordered in order to pour the entire sidewalk. Explain.

(b) If the sidewalk is to be 1 meter in width, determine the amount of gravel that has to be ordered to lay under the entire sidewalk. Explain.

(c) If the sidewalk is n meters in width, determine a general model (analytic formula) that will give the amount of cement that has to be ordered in order to pour the entire sidewalk. Explain.

(d) If the sidewalk is n meters in width, determine a general model (analytic formula) that will give the amount of gravel that has to be ordered to lay under the entire sidewalk. Explain.

(e) Determine ways in which this model could be in error.

8. A tree ten meters away from the front of a house is to be cut down. The tree surgeon notices a 3 meter high lamp post that has a shadow of about 1.2 meters. She then measures the tree shadow to be about 3.9 meters in length.

(a) Estimate the height of the tree.

(b) Explain if there is a possibility that the tree will hit the house.

(c) Determine ways in which this model could be in error.

9. An alien species of fish has been introduced into a local farmer's lake which serves as a water supply for her dairy cows. Her concern is that the fish may somehow contaminate the lake thus rendering it useless as a water resource for her cows. She wants to know something about the speed at which the fish are reproducing. The following graph 1.3 represents the population growth of the fish over a period of time (population vs. time). Time is measured in years.

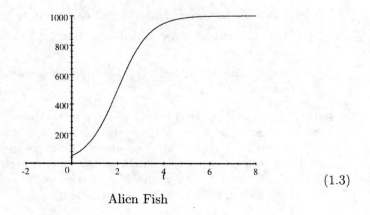

(1.3)

Alien Fish

(a) Approximately, how many fish were initially introduced into the lake? Explain.

(b) Approximately, how many fish can the lake sustain? Explain.

(c) Approximately, how many years will it take for the population to reach its maximum number? Explain.

(d) In what years is the fish population growing the fastest? slowest? Explain.

10. The graph 1.4 below shows the temperature pattern over a 48-hour period. Use this information to answer each of the questions below.

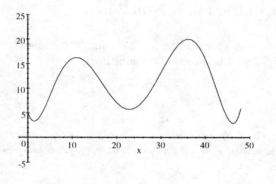

(1.4)

Two-Day Weather Pattern

(a) What were the maximum and minimum temperatures over the first 24 hours?

(b) What were the maximum and minimum temperatures over the last 24 hours?

(c) What were the maximum and minimum temperatures over the entire 48 hours?

(d) What was the range in temperature over the 48 hours?

(e) What was the fastest rise in temperature over the 48 hours?

(f) What was the fastest drop in temperature over the 48 hours?

1.3   Role of a Function in Models

The concept of a mathematical function plays an important role in the construction of all three types of models studied in this text: geometric, analytical and numerical. Before we continue with the development of mathematical models we will, therefore, spend some time reviewing the concept of a function. This section discusses the concept of a function from a modeling perspective. Several examples of the role of functions in the construction of models are included.

1.3.1   Definition of a Function

Ever since we first studied algebra as young students we have been exposed to the definition of a function. The definition as it appears in many algebra texts is

Definition 13 *A **function** is a rule which assigns to each element of one set, called the **domain**, exactly one element from a second set, called the **range** or **co-domain**.*

This definition has little meaning for students. It is generally presented as an abstract mathematical concept and students, rightly, view it as remote to their interests. So, why is it presented so prominently in mathematics texts? In fact, the concept of a function is so important in mathematics that within conversations among mathematicians the concept is never mentioned. That is, the concept of a function is so basic to the use and application of mathematics that it is assumed every mathematician already has a clear, intuitive understanding of what a function is. Is there some way, then, that we students can come to a better understanding of the concept of a function? Perhaps a different approach will be helpful.

In the section 1.2.1 we discussed the example of tracking the spread of a cold through an office building given certain conditions of frequency of contact with others, resistance to germs, etc. Recall that you used both the table 4 and the graph in exercise 7 to answer certain questions about the rate at which the cold was spreading. In doing so notice that there was a relationship between the time of day and the total number of people infected. That is, for each time of day there was a specific number of infected people associated with that particular time of day. To know the time of day is to know how many people are infected. Let's look at other examples where this type of association takes place. First we will consider the following exercise.

Exercise 14 *In each supermarket manager's office is kept a master list of the food items for sale along with the price that is assigned to each item. Sometimes a copy of this list is kept at each register so that the cashier can quickly check the price of an unknown item. For example,*

$$apples.............\$1.29/lb$$
$$lettuce.............\$0.99/head$$
$$cereal.............\$3.29/15oz$$

$$\vdots \qquad\qquad \vdots$$

Of course the list assigns a price to every item of food and it assigns exactly one price to each item of food.

1. *Explain why it is important (to store operations) that every food item be assigned a price. That is, what would happen if a food item on the list did not have a price listed next to it?*

2. *Explain why it is important (to store operations) that every food item be assigned exactly one price. That is, what would happen if a food item on the list had two or more prices listed next to it?*

3. *Every food item is assigned a price so that the supermarket will operate smoothly. Is it necessary that every available price be used as well in order for the supermarket to operate smoothly? Explain.*

Here is another example of a simple but necessary assignment.

Exercise 15 *In radar towers at airports a daily list is kept of the airplanes scheduled to fly that day and the takeoff times assigned to those planes. This, of course, allows air traffic controllers to keep track of airport traffic. For example,*

$$NorthAir\ Flight\ 445............07{:}05h$$
$$SouthAir\ Flight\ 839............10{:}15h$$
$$EastAir\ Flight\ 309.............13{:}30h$$

$$\vdots \qquad\qquad\qquad \vdots$$

Once again a takeoff time is assigned to every flight that day and each flight scheduled to takeoff is assigned exactly one takeoff time.

1. *Explain why it is important (to both customers and airlines) that every flight scheduled to fly that day is assigned a takeoff time. That is, what would happen if a flight scheduled to fly was not assigned a takeoff time?*

2. *Explain why it is important (to both customers and airlines) that every flight scheduled to fly that day is assigned exactly one takeoff time. What would happen if a flight scheduled to fly was assigned more than one takeoff time?*

3. *Every flight scheduled to fly that day is assigned a takeoff time so that airport operations proceed smoothly. Is it necessary that every available takeoff time be used as well in order for airport operations to proceed smoothly? Explain.*

The two previous examples demonstrate the importance of carefully and systematically assigning objects from one set (*food* in the first example, *airplanes* in the second example) to elements from another set (*prices* in the first example, *takeoff times* in the second example). In each example above carefully assigning food to specific prices or planes to takeoff times allows each system to function smoothly. Without this assignment method there would be chaos and many angry customers. Of course, the student should recognize at this point that this assignment process is exactly the definition of a function as stated in the beginning of this section. This method of assigning elements from one set to elements from another set also plays a significant role in mathematics. This assignment concept helps to organize and clearly identify solutions to many problems. Note that in the previous section when we were modeling the spread of the cold each time of day (the *domain*) is assigned, through our calculations, exactly one number of total infected people (the *range*).

These examples illustrate a more general definition of a function which is more appropriate for our purposes.

Definition 16 *A function is a pairing of the elements of one set, called the domain, with the elements of a second set, called the co-domain or range, that satisfies the following conditions:*

1. *Every element of the domain is paired with an element of the range.*

2. *Each domain element is paired with one range (co-domain) element.*

The store manager or air traffic controller are not thinking to themselves "I am using a function," nor do they care. They only want operations to run smoothly. Nevertheless, a function is in fact what they are using to keep things running smoothly. On the other hand, mathematicians may not be interested in the basic operations of supermarkets or airports. The power of mathematics is that the same mathematical concept (in this case a function) can be applied to two completely different areas (in this case supermarkets and airports). That is why many mathematicians console themselves with the study of purely numerical (abstract) examples.

1.3.2 Notation: A Purely Numerical Example

We will spend only a brief moment studying purely numerical examples of functions. This is necessary to see how to better organize data which often appears as a sequence of numbers and to recognize the proper notation that is used. So what is a sequence of numbers?

Definition 17 *A **sequence of numbers**, for example $\{5, 9, 13, 17, 21, 25, \dots\}$, will be defined here as a function whose domain is the set of natural numbers $\{1, 2, 3, 4, 5, \dots\}$, and whose co-domain is the sequence of numbers itself, for example $\{5, 9, 13, 17, 21, 25, \dots\}$.*

That is, each natural number represents each corresponding "position" in the sequence. Thus, the assignment is "position n" is assigned the nth number in the list.

$$
\begin{array}{lll}
\text{position 1} & \text{is assigned} & 5 \\
\text{position 2} & \text{is assigned} & 9 \\
\text{position 3} & \text{is assigned} & 13 \\
\text{etc.} & \text{etc.} & \text{etc.}
\end{array}
$$

In many cases it is helpful to find a general expression that represents the sequence or function. It is easier to write one expression than an entire list of numbers. Here are some examples of sequences where we find a general functional expression to represent each sequence. Note that the domain of natural numbers is "invisible" since it is understood that the domain is the "position" of each co-domain element. We will consider the following sequences and determine a functional expression that represents each.

1. $\{4, 8, 12, 16, 20, 24, \ldots\}$

2. $\{5, 9, 13, 17, 21, 25, \ldots\}$

3. $\{\frac{1}{5}, \frac{1}{9}, \frac{1}{13}, \frac{1}{17}, \frac{1}{21}, \frac{1}{25}, \ldots\}$

4. $\{-\frac{1}{5}, \frac{1}{9}, -\frac{1}{13}, \frac{1}{17}, -\frac{1}{21}, \frac{1}{25}, \ldots\}$

Solution 1. Successive elements of the sequence $\{4, 8, 12, 16, 20, 24, \ldots\}$ have a difference of 4. That is, this sequence represents the positive multiples of 4 ($f(1) = 4, f(2) = 8, f(3) = 12, \ldots$), that is, 1 is paired with 4, 2 with 8, etc. Therefore, $f(n) = 4n$, n a natural number, is a function expression for the sequence.

Solution 2. Successive elements of the sequence $\{5, 9, 13, 17, 21, 25, \ldots\}$ have a difference of 4. Again it seems this sequence represents the positive multiples of 4 except that each entry is off by one ($f(1) = 5, f(2) = 9, f(3) = 13, \ldots$). Therefore, $f(n) = 4n + 1$, n a natural number, is a function expression for the sequence.

Solution 3. By comparing the sequence $\{\frac{1}{5}, \frac{1}{9}, \frac{1}{13}, \frac{1}{17}, \frac{1}{21}, \frac{1}{25}, \ldots\}$ with the sequence in the previous example (example 2) we see that each entry in this sequence is the reciprocal of each corresponding entry in the previous sequence. Hence, $f(n) = \frac{1}{4n+1}$, n a natural number, is a function expression for the sequence.

Solution 4. Again by comparing the sequence $\{-\frac{1}{5}, \frac{1}{9}, -\frac{1}{13}, \frac{1}{17}, -\frac{1}{21}, \frac{1}{25}, \ldots\}$ with the sequence in the previous example (example 3) we see that the only difference is this sequence alternates in sign (the odd terms are negative and the even terms are positive). Hence, $f(n) = \frac{(-1)^n}{4n+1}$, n a natural number, is a function expression for the sequence.

Notice that in each example above we are using the notation $f(n)$ to represent the functional expression. In general, the "f" part represents the "name" of the function where as the variable "n" represents the domain quantities. Together, $f(n)$, read "f of n," represents the range/co-domain quantity. Usually, in pure math classes $f(n)$ or $f(x)$ is sufficient for representing functions. However, when applying mathematics to models it more appropriate to pick letters that will remind us of the application being used. For example, if we want to model how fast a rocket is moving over time we might represent the functional expression as $S(t)$ since S reminds us of speed and t reminds us of time better than f and x would.

1.3.3 Measuring the Severity of a Hurricane: An Application

Here is an example of how such sequences and assignments are used in everyday experiences. The table 1.5 below refers to various wind velocities and the resulting forces generated in pounds per square foot. This would be of interest to a meteorologist since the force generated by wind is usually used a means of measuring the severity of an upcoming storm: the greater the velocity, the greater the force applied by the wind, the greater the damage it might incur. Note that table 1.5 has several sequences. The first row is simply the domain, that is, the "position" indicator. Although it was not included in the previous "purely numerical" examples, it is included here for completeness and as a guide through the example. Eventually, as we proceed through the text, we will be interested in finding a general "assignment" expression for each sequence. This would help the meteorologist in the sense that instead of generating the table each time she would only have to refer to the general expression to determine a specific velocity. For the purposes of this example, however, we will concentrate only on finding a general functional expression for the first sequence, *wind velocity*. The hunt for a general functional expression for the Force will be considered in a later chapter.

Exercise 18 *(This exercise could also be done with the aid of a calculator or spreadsheet.) Acting as a meteorologist you have set up equipment designed to measure the velocity of the wind and the force it generates during a recent storm. This information is recorded in 1.5 below.*

Term(n)	1	2	3	4	5	6	7	8	9
Wind Velocity($W(n)$)	30	35	40	45	50	55	60	65	70
Force(lb/sq ft)	4.5	6.125	8.0	10.125	12.5	15.125	18.0	21.125	24.5

$$(1.5)$$

1. *Write the wind velocity (W(n)) as a separate sequence of numbers using the "braces" notation found in section 1.3.2.*

2. *By following the examples from section 1.3.2, find a general function expression, W(n), for the wind's velocity.*

3. *Use this general function expression to determine the 15th term of this sequence. the 20th term of this sequence. the 25th term of this sequence.*

4. *Explain what some of the advantages are to using the general function expression to find addition terms in the sequence rather than simply continuing to extend the list of numbers.*

5. *Storms that originate in the Atlantic ocean are not classified as hurricanes until their winds reach 75 miles per hour. As a meteorologist you would be most interested in the force generated by winds that are in excess of 75 miles per hour. Give some reasonable explanation as to why your data stops at the ninth term of the sequence (70 miles per hour).*

6. *Plot the wind velocity sequence on the grid below. Label the vertical axis as WIND VELOCITY and the horizontal axis as TERM. Before marking off the numbers on both axes think about which range of numbers would be appropriate to appear on the axes.*

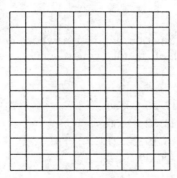

7. *Sketch in a curve that best fits the data points. Describe what type of curve it is. Explain how you could use the graph to answer question 3 in this exercise.*

A different technique is required for finding the general function expression for the force generated by wind velocity, the second "force" sequence in table 1.5. This technique will be covered in Chapter 3.

1.3.4 Problems

1. In section 1.3.1 we described functional assignments using a supermarket and airplane traffic as examples. Describe up to three (3) other examples in the world where it is important that each element from the first set is assigned to exactly one element from the second set. For your examples explain why it is important that this type of assignment take place. (Note: Several such examples can be found around campus.). Clearly state which set is the domain and which is the range.

   (a) Example:

   Domain:

   Range:

   (b) Example:

   Domain:

   Range:

   (c) Example:

   Domain:

   Range:

2. Find a general function expression whose domain is the set of natural numbers for each of the following sequences.

   (a) Linear sequences:

       i. $\{5, 11, 17, 23, 29, 35, \ldots\}$

       ii. $\{5, 14, 23, 32, 41, 50, \ldots\}$

       iii. $\{7, 11, 15, 19, 23, 27, \ldots\}$

       iv. $\{13, 21, 29, 37, 45, 53, \ldots\}$

       v. $\{27, 23, 19, 15, 11, \ldots\}$

       vi. $\{18, 11, 4, -3, -10, \ldots\}$

   (b) Rational sequences:

       i. $\{\frac{1}{3}, \frac{1}{7}, \frac{1}{11}, \frac{1}{15}, \frac{1}{19}, \frac{1}{23}, \ldots\}$

       ii. $\{\frac{1}{5}, \frac{1}{7}, \frac{1}{9}, \frac{1}{11}, \frac{1}{13}, \frac{1}{15}, \ldots\}$

       iii. $\{\frac{1}{3}, \frac{5}{3}, \frac{9}{3}, \frac{13}{3}, \frac{17}{3}, \frac{21}{3}, \ldots\}$

       iv. $\{\frac{1}{2}, \frac{2}{3}, \frac{3}{4}, \frac{4}{5}, \frac{5}{6}, \frac{6}{7}, \frac{7}{8}, \ldots\}$

       v. $\{\frac{3}{7}, \frac{6}{10}, \frac{9}{13}, \frac{12}{16}, \frac{15}{19}, \frac{18}{22}, \frac{21}{25}, \ldots\}$

   (c) Alternating sequences:

       i. $\{-7, 12, -17, 22, -27, 32, \ldots\}$

       ii. $\{-9, 16, -23, 30, -37, 44, \ldots\}$

       iii. $\{-7, 11, -15, 19, -23, 27, \ldots\}$

       iv. $\{-13, 21, -29, 37, -45, 53, \ldots\}$

3. Turnpikes in each state have a commission whose job it is to over see turnpike operations such as travel fees and speed limits, among other things. A local state turnpike commission has determined that in order to cover turnpike operating and repair costs the toll to individual automobiles will be a base rate of $0.50 plus an additional $0.04 per mile for every mile travelled. The total fee is paid at a toll booth as the motorist exits the turnpike.

   (a) Fill in the table below with the corresponding cost to each motorist for the first few miles of travel on the turnpike

Miles Travelled(n)	1	2	3	4	5	6	7
Toll Cost ($T(n)$)							

   (b) Find a general function expression for the Toll Cost, ($T(n)$), for travelling on the turnpike.

   (c) Use your answer in part b. to determine the toll for a trip on the turnpike that is

      i. 100 miles
      ii. 150 miles
      iii. 200 miles

   (d) Use your answer in part b. to determine your turnpike cost if you were travelling from...

      i. mile marker 256 to mile marker 334.
      ii. mile marker 256 to mile marker 31.
      iii. mile marker 31 to mile marker 334.

   (e) It has been stated that most models are simply approximations and therefore have some built-in error. Is that true in this example. If so, describe some ways in which this model could be in error. As a member of the commission how would you handle such error?

4. Small, engineering firms provide a unique service to larger companies. Often when a larger design firm is unable to meet demand or is unequipped to provide a service the company will contract out part of its work load to the smaller firm. One such small firm was contracted to design rectangular plastic pin housings for electrical connections. These are rectangular pieces of plastic with "holes" in them (see diagram 1.6). Metal pins are placed in specific holes to connect one piece of hardware to another. The dimensions of these housings need to be of different sizes depending on their use. For reasons of stability, electrical current, etc. the rows of pin holes needed to be one millimeter in diameter and 2 millimeters apart; that is, 3 millimeters all together were needed for each row of pin hole placement. So, if three rows of pin holes were required the housing had

to be 11 millimeters high (Why 11 millimeters and not 9 millimeters?). The columns of pin holes needed to be one millimeter in diameter and 3 millimeters apart. For example, if six columns of pin holes were required the housing had to be 27 millimeters long.

$$
\text{Height}
\begin{array}{|cccccc|}
\hline
\text{O} & \text{O} & \text{O} & \text{O} & \text{O} & \text{O} \\
\text{O} & \text{O} & \text{O} & \text{O} & \text{O} & \text{O} \\
\text{O} & \text{O} & \text{O} & \text{O} & \text{O} & \text{O} \\
\hline
\end{array}
\qquad (1.6)
$$
$$\text{L e n g t h}$$

(a) Fill in the table below with the corresponding height for each housing having the indicated number of rows.

Number of Rows (n)	1	2	3	4	5	6
Housing Height $(H(n))$						

(b) Find a general function expression for the Housing Height, $(H(n))$.

(c) Use your answer in part b. to determine the height of a housing that has...

  i. 10 rows

  ii. 20 rows

  iii. 30 rows

(d) Fill in the table below with the corresponding length for each housing having the indicated number of columns.

Columns on Housing(n)	1	2	3	4	5	6
Housing Length $(L(n))$						

(e) Find a general function expression for the Housing Length, $(L(n))$.

(f) Use your answer in part b. to determine the width of a housing that has...

  i. 10 columns

  ii. 20 columns

  iii. 30 column

(g) Describe some ways in which this model could be in error.

5. Today it is possible to rent everything from houses to furniture to TV's to cars. Rental companies offer a variety of deals to its customers, including an option to buy the object at the end of the rental lease. Most such deals simply apply each month's rent, less the security deposit, to the purchase price of the item. This is referred to as a *lease to buy* option. Such operations as medical groups, travel agencies, etc. use the lease to buy

option for office furniture. You decide to go into business for yourself and intend to use the lease to buy option for your office furniture. The rental company offers your a lease that requires a $200 security deposit plus a rental fee of $75 per month. The $200 security deposit is also applied toward the purchase price of the furniture.

(a) Fill in the table below with the corresponding (total) cost to rent the furniture for the corresponding number of months.

Number of Months (n)	1	2	3	4	5
Total Rental Cost ($C(n)$)					

(b) Find a general function expression for the Total Rental Cost, ($C(n)$).

(c) Use your answer in part b. to determine the total cost to rent the furniture for...

   i. 10 months
   ii. 2 years
   iii. 3.5 years

(d) Use your answer in part b. to determine how many months it would take to pay off (and thus own) the furniture if the security deposit is applied toward the purchase price of the furniture and if the furniture was priced at...

   i. $2400.
   ii. $5000.
   iii. $7500.

(e) Use your answer in part b. to determine how many months it would take to pay off (and thus own) the furniture if the security deposit **is not** applied toward the purchase price of the furniture and if the furniture was priced at...

   i. $2400.
   ii. $5000.
   iii. $7500.

(f) It has been stated that most models are simply approximations and therefore have some built-in error. Is that true in this example. If so, describe some ways in which this model could be in error. How would you handle such error?

6. Although watching giant insects attack earth is a popular topic of science fiction movies, it is mathematically impossible for insects to grow those dimensions. As biologists well know, there is a limit to how much "mass"

(3-dimensional volume) can be supported by the "feet" (2-dimensional surface area) of the creature. This exercise demonstrates how to relate area (2-dimensions) to volume (3-dimensions). We will consider the simple experiment of "stacking" cubes. Diagram 1.7 shows a single 1cm by 1cm by 1cm cube and a 2cm by 2cm by 2cm cube (or eight 1cm by 1cm by 1cm cubes). Similarly we could construct a 3cm by 3cm by 3cm cube, a 4cm by 4cm by 4cm cube, etc.

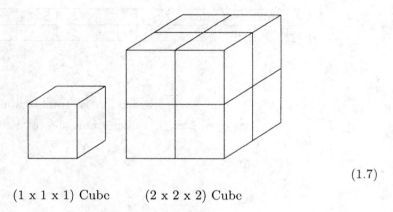

(1.7)

(1 x 1 x 1) Cube   (2 x 2 x 2) Cube

(a) Fill in the table below with the corresponding surface area and volume for each cube dimension.

Dimension of Cube (n)	1	2	3	4	5	6
Sruface Area ($A = A(n)$)	$1cm^2$	$24cm^2$				
Volume of Cube ($V(n)$)	$1cm^3$	$8cm^3$				

(b) Find a general function expression for the volume of the cube, $V(n)$, in terms of the surface area of the cube, A. That is, surface area is the domain and volume is the range.

(c) Use your answer in part b. to determine the volume of a cube of dimension...

   i. 10cm

  ii. 25cm

 iii. 4.6cm

1.4 Summary

A mathematical model is the abstract equivalent of such concrete models as toys and maps. A mathematical model uses mathematical concepts such as equations, functions, graphs, charts, etc. to represent the important characteristics of a real-world problem. Every model looses details from the original problem. As a result error is a natural part of the process of constructing models. A good model is one that maintains the characteristics that we are interested in studying as well as minimizes, or at least recognizes, the possibility of error in the results.

There are primarily three types of mathematical models: geometrical, analytical, and numerical. All three have advantages and disadvantages to representing and solving real-world problems. As a result we often try to create a model of a problem that utilizes more than one type of mathematical model.

However, no matter which type or combination of types of models is used, the concept of a function is central to all of them. We may not realize it but we use the concept of a function for many everyday operations. The reason for this is that the function concept bring order to a system so that the system will operate smoothly. This is true in pure mathematics as well as applied mathematics. The function brings order to mathematics as well as the real world.

In future chapters we will expand on these concepts. Specific functional models and their applications are studied in detail in each chapter.

Chapter 2

Linear Function Models

2.1   Introduction

In recent years the several news items have discussed the increase in the number of multiple births in the United States. This, at least in part, is due the increased availability of fertility drugs. One such article is reproduced in Chapter 5. This article reported the rate at which the number of multiple births has been increasing over the previous twenty years. We will discuss the mathematics used in that article in Chapter 5. For now, however, what this report represents is the importance of knowing how and why the number of multiple births is changing.

Doctors, scientists and business personnel are very interested in determining the rate at which things change over time. Knowing how things change allows us to plan for the future. As in the "multiple births" article, if we know the rate at which children are being born this allows us to determine the how many schools will be needed in the near future; how many families will need housing, cars, food, etc.; how much waste will be generated and so how many incinerators, landfills and other waste disposal areas will be needed; what energy consumption will be; etc. Predicting and planning for the future is of major concern too that compares how one quantity changes with respect to another quantity. For example, crime statistics are often reported as ratios: One person is mugged every 30 minutes in the United States. This does not mean that criminals wait for 30 minutes before they mug someone else. It simply means that "on average" one person is mugged every 30 minutes. Ratios are meaningless without units. Always remember to include the units when working with ratios. The mugging example compares number of people to time. There are many other examples of ratios, number of children per family, number of births per year, income per family, etc. One of the most common rates of change is that of speed. Speed is defined as the many professionals in all fields of study. In this section we will be primarily concerned with linear models, one of the simplest models of change.

2.2   Rates of Change

2.2.1   Speed

Speed is defined as the rate at which distance is traveled over time. Since the invention of cars children have been asking:" Are we there yet?" With a knowledge of ratios and rates of change children could easily determine "when we get there." We will begin our study of rates of change with the concept of speed. In the box below write down the formula that relates distance traveled to rate and time.

Exercise 19 *With modern technology it is possible for distribution companies to "keep track" of their trucks electronically. Each truck is equipped with a signaling devise that sends out periodic signals to satellites. These signals are then bounced off the satellite and back to the dispatcher's office where a truck's location on the highway is then determined. As a dispatcher you are particularly interested in the trip one of your drivers is making from Harrisburg (company headquarters) to Washington, D.C. (a distribution site for the company). At various time intervals, the signal indicates the location of a truck's distance from Harrisburg. The entire trip runs 105 miles. Refer to the map below for details. The times recorded are listed in the following table.*

Point	Time Travelled HH:MM	Distance Travelled MILES
A	0:00	0
B	0:15	10
C	0:30	10
D	1:00	45
E	1:45	90
F	2:15	90
G	2:45	105

1. *Sketch a graph of your entire trip showing total distance (range) travelled versus time (domain) and label each point according to the labels given in*

the table.

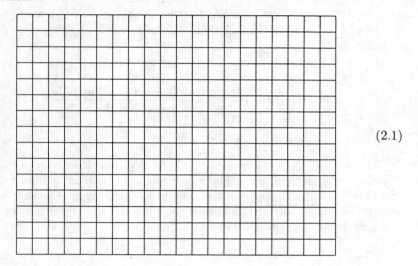

$$(2.1)$$

2. *What was the driver's average speed between points A and B? D and E? between points F and G? (Note that when computing the speed (velocity) of the truck, the times are given in minutes. However, we usually do not record highway speeds in miles per minute or feet per minute but in miles per hour. Be sure to properly convert your speeds to the appropriate units for discussion.)*

3. *Assuming she left the house at 9:00 AM, explain what may have been happening between points B and C and between points E and F.*

4. *Assuming she was on the interstate throughout the entire trip, explain what may have been happening between points A and B, between points D and E and between points F and G.*

In each case the speed is calculated by determining the distance travelled and dividing by the time elapsed. That is, since each pair of points on the graph 2.1 represents another reading then the average speed between each such pair of points is

$$speed = \frac{\text{distance travelled between the points}}{\text{time elapsed between the points}}. \qquad (2.2)$$

The *y-coordinates* of each pair of points determine the distance the truck has traveled between the points. The *x-coordinates* of each pair of points determine the time that has elapsed between the points. Since the distance travelled is marked on the graph vertically and the time elapsed is marked on the graph horizontally we can rewrite formula 2.2 as

$$\text{rate of change (speed)} = \frac{\text{vertical change from } 1^{st} \text{ point to } 2^{nd} \text{ point}}{\text{horizontal change from } 1^{st} \text{ point to } 2^{nd} \text{ point}}. \qquad (2.3)$$

This is true for every pair of consecutive points. We would like to write a general expression that models what we are doing in order to calculate the speed. It is inefficient to keep writing similar calculations for each consecutive pair of points. It is more efficient to develop an analytical expression that represents all such computations. Therefore, if the first point used is identified as (x_1, y_1) and the second point used is identified as (x_2, y_2) then formula 2.3 can be written mathematically as

$$\text{rate of change} = \frac{y_2 - y_1}{x_2 - x_1}. \qquad (2.4)$$

When the value in formula 2.4 is associated with the line from which the points have been taken this value is referred to as the **slope of the line**. We will have more to say on this later. Before we officially define the slope of a line, let's return to the truck example. Most distribution companies do not have a single truck. They have an entire fleet of trucks, usually several trucks travelling the same route. In the previous exercise you were trying to predict what the driver was doing between points by inspecting the speed of the truck. It is difficult, if not impossible, to predict exactly what the driver was doing simply because knowing the driver's speed in not enough information. The dispatcher can usually get a better idea as to what the driver's behavior is by comparing their behavior with another driver's behavior who is travelling the same route.

Exercise 20 *You monitor a second truck that made the same trip along the same route at the same time as the first truck. Both trucks start out at the same time. The second truck's information is recorded below.*

Truck 2 Point	Time Travelled HH:MM	Distance Travelled MILES
A	0:00	0
B	0:30	20
C	1:00	50
D	1:30	50
E	2:00	90
F	2:15	105

1. *Plot Truck 2's data on the same graph as the first truck. Analyze the trip made for truck 2 as you did for truck 1 above.*

2. *Which truck arrived in Washington first? Explain how you know this.*

3. *Did one truck pass the other along the way? Which truck passed the other? When and where? Explain how you know this.*

The graphs represented by both trucks above are referred to as **piecewise graphs** because they are composed of several pieces of linear graphs. Piecewise graphs will be studied in more depth in later sections.

2.2.2 Problems

1. Here are a few more questions concerning the truck distribution examples graphed in 2.1.

   (a) If the first driver was to maintain the same speed throughout the rest of the trip as she had between points D and F, when would she have arrived in Washington?

   (b) What was her average speed over the course of the entire trip?

   (c) What was the average speed of the second truck between points B and D?

   (d) Which truck made better time in the first hour of the trip? in the last hour of the trip?

   (e) If each truck had maintained its original speed (between points A and B) throughout the entire trip without stopping which one would have arrived in Washington, D.C. first?

2. One state has set a sales tax of 6%. This percentage represents a ratio.

   (a) Describe the ratio as to what is meant by a 6% sales tax.

   (b) What is the tax on $54.24 in taxable goods in this state?

   (c) If you paid $6.02 in taxes on taxable goods, how much did you spend on goods?

   (d) If your total bill came to $63.78 how much of your total was in taxes?

3. Recently a lunar probe discovered "ice" on the moon. Upper estimates of the amount of lunar ice is enough water to fill a 4 square mile lake 35 feet deep but that the water is spread out over a total of 25,000 square miles at both the north and south poles. Naturally, this "water" is not just sitting on the surface of the moon, it is embedded into moon's soil.

   (a) How much water would fill a lake of dimensions a 4 square mile lake and 35 feet deep?

   (b) Write a ratio that identifies the amount of water that can be found per square mile on the moon.

   (c) Suppose that the space agency is interested in establishing a lunar science station that is to cover 1000 square miles. Use your answer in part b. to determine approximately how much water the scientists can expect to find.

4. Most schools are too large to collect data by asking each and every student. Conduct a "survey" of students and use to answer each question.

   (a) What is the ratio of left-handed students on campus to the size of the total student body?

(b) What is the ratio of students who take advantage of a university food service plan to those who do not?

(c) What is the ratio of the number of students enrolled in a mathematics class this semester to the total student body?

(d) What is the ratio of the number of students who live on campus to those that live off campus?

(e) What is the ratio of the number of out-of-state students to the number of in-state students?

(f) Write ratios that determines the political affiliation of the student body.

5. Choose a building on campus.

(a) Measure the length of the shadow cast by the building. Measure your height and the shadow that you cast. Use this information to determine the height of the building.

(b) Use this procedure to determine the height of the tallest tree on campus.

(c) Shadows can only be measured if the sun is present. Also, the building's shadow may be too long to measure accurately. Determine another means by which ratio and proportions can be used to measure the height of the building (without the use of shadows).

6. A recent report on tourism stated that the number of tourists is on the rise across several different sections of the United States. Although we all love to "get away from it all" from time to time this is not necessarily good news for the over-burdened public parks and monuments across the country. The report stated that in the Middle Atlantic States tourism was up 9% from the 12 months from January, 1997 to December, 1997.

(a) If the number of tourists to the mid-Atlantic states in the first month was 5,400,000 find a linear model that represents a steady 9% growth in tourism over the previous 12 months.

(b) Use this information to determine the number of tourists in the calendar year of 1996 (the previous 12 months).

2.3 Linear Rates of Change

When dealing with linear models the analytic representation ($y = mx+b$) always exists and is very easy to find. All that is required to obtain the linear analytic model of a problem is two points from the model. These points are easily read off the graph, if given, or from a numerical table of data points. Therefore, in this section we will concentrate on obtaining the analytic representation first from a graph (section 2.3.1) and then from a numerical table (section 2.3.2).

2.3.1 Geometric and Analytic Representations of Linear Models

As mentioned in section 2.2 the slope of a line represents the rate at which the vertical (range) item is changing in comparison to how the horizontal (domain) is changing. Recognizing the (analytical) formula for slope as a measure of the slant of the (geometrical) line demonstrates the first connection between analytical and geometric models. Using more than one type of model simultaneously gives a better representation of the problem under consideration. Connections between geometric, analytical and numerical models will continue to be stressed throughout this text. Several more examples are presented here to once again demonstrate the relationship between the concept of rate of change and slope of a line.

Exercise 21 *The graph 2.5 represents the amount of oil pumped from an oil well each year over its production life. The domain is in years and the range is in billions of barrels of oil.*

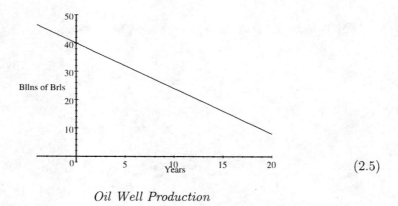

(2.5)

Oil Well Production

1. *How much oil was in the well at the beginning of the well's production? Write your answer as an ordered pair. Explain.*

2. *Estimate how long will it take for the oil well to run dry? Write your answer as an ordered pair. Explain. (Hint: carefully extend the x-axis and the given line.)*

3. *At what rate(billions of barrels per year) is the oil being pumped out of the well? Label the rate clearly and show all calculations. (Note that the calculation of the rate is the slope of the line in 2.5.)*

4. *Your answer for the rate in part 3. should be negative. Why? Explain.*

5. *If there are 20 billion barrels of oil left in the well this year (1998) when*

will the well run dry? Explain.

In this example we were able to gain a great deal of information from the given graph. There are many instances in which a graph serves very practical purposes, for example, giving a presentation to company executives, fellow students, etc. A picture goes a long way in explaining ideas. However, graphs and charts are inexact and there are times when something more accurate is needed. When dealing with linear models finding an *analytic representation* of the model, that is, an algebraic equation, is quite simple.

Recall from past Algebra classes the following.

Definition 22 *The **slope-intercept form of a line** with slope m and y-intercept $(0, b)$ is*

$$y = mx + b \ (or \ f(x) = mx + b). \tag{2.6}$$

Definition 23 *The **point-slope form of a line** with slope m and (x_0, y_0) a point on the line is*

$$y - y_0 = m(x - x_0). \tag{2.7}$$

Exercise 24 *Find an equation $y = mx + b$ (or $y - y_0 = m(x - x_0)$) that represents the line in the oil well graph above.*

Here is a similar example.

Exercise 25 *Recent Statistics show that the use of cellular phones is one of the fastest growing industries world wide over the previous year. The graph 2.8 represents the growth in the number of subscribers to cellular phone service over the 12 months from January, 1997 (t = 0) to December, 1997 (t = 12). The domain is in months and the range is in millions of subscribers.*

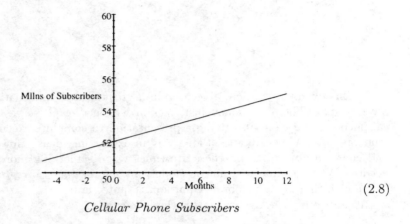

(2.8)

Cellular Phone Subscribers

1. *How many subscribers were there at the beginning of the year? Write your answer as an ordered pair. Explain.*

2. *How many subscribers were there at the end of the year? Write your answer as an ordered pair. Explain.*

3. At what rate were subscribers growing over this year? Label the rate clearly and show all calculations.

4. Write the equation of the line that represents modeling the number of subscribers over the last year.

5. Use the analytic model in part 4 above to determine projected subscribers 3 months from now; 18 months from now. Of the two projections you just figured out, which do you think is more accurate? Why?

Note that in the oil well exercise the slope (rate at which oil was being pumped out of the well) was a *negative ratio* and in the cellular phone subscriber exercise the slope (rate of new subscribers) was a *positive ratio*. This coincides with the graphical representation of each exercise. We "read" a graph the same way in which we read text, from left to right. In that case, the "oil well production" line is falling (decreasing), an indication that its slope should be negative, whereas the "cellular phone subscriber" line is rising (increasing), an indication that the slope should be positive.

2.3.2   Numerical and Analytic Representations of Linear Models

In the previous section 2.3.1 we obtained the analytic model from data that was read from a given graph. As helpful as a graph can be, it is not always immediately available for us to use nor is it very accurate. In this section we continue our discussion of linear models, however, instead of developing the analytic representation from a graph we will develop the analytic representation from data that is generated in table form. That is, we will develop the connect between a numerical representation (a table of data) and the analytic representation (equation) of a linear model.

Exercise 26 *Cellular phone companies have an "up-front" surcharge for each call that is made in addition to the per minute charge. That is, customers are charged for "making the connection" as well as the per minute charge. A local service, which we will refer to as Service A, offers its customers a $3.00 surcharge along with a $0.16 per minute charge for each call they make.*

1. Fill in the table entries for each length of phone call.

Length of call (in min.)	5	10	15	20	25
Service A					

(2.9)

2. *Plot each point on the grid provided and label clearly. Identify the y-intercept of the line that passes through these points. What do you notice about the pattern of points on the graph?*

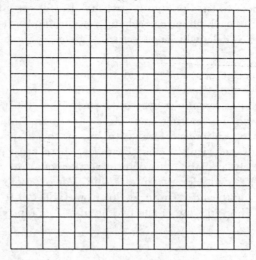

3. *Use two values from the table 2.9 to find the slope of the line in the graph above. Choose two different values from table 2.9 and calculate the slope of the line again. What can you conclude from this?*

4. *Determine the equation of this line. (The choice of method is yours) What do the variables x and y represent in this equation?*

5. *What is the relationship between the slope of the line and the per minute rate the cellular company charges?*

Clearly the per minute rate of $0.16 per minute that the phone company charges is the same as the slope of the line in the graph. Once again the slope of the line is a ratio quantity that measures the rate at which the vertical quantity varies with respect to the horizontal quantity. In this case, the vertical quantity (the dollar value of each phone call) changes according to the horizontal quantity (the length of each call) which, of course, is given as $0.16 per minute.

The **y-intercept** of a linear model is defined as the point where the line intersects the y-axis. Clearly, this is the point where the value of $x = 0$. In each of the previous examples we could easily find the point $(0, b)$ either by reading the graph or choosing the data point in the table where $x = 0$. In each case the slope-intercept form of a line is the easiest form to use when determining the analytical model. However, there are times when this point is not immediately available. To determine the equation of a line in that case we would use the point-slope form of a line. For example, suppose we use the points $(10\,\text{min}, \$4.60)$ and $(20\,\text{min}, \$6.20)$ from table 2.9 to find the equation of the line. Then, choosing one of the two points, the point-slope form of the line yields

$$y - 4.60 = 0.16(x - 10)$$

which simplifies to

$$y - 4.60 = 0.16x - 1.60)$$

or

$$y = 0.16x + 3.00.$$

Use the point-slope form of a line when working the next example.

Exercise 27 *You recently acquired a local truck rental franchise that rents 12-foot trucks. After assessing the dynamics of the market and your possible costs you decide that a fair rental fee for the 12 ft trucks is to charge customers a flat $40.00 surcharge up front plus $0.31 per mile. Use this information to answer each of the following.*

1. *Fill in the table entries for the cost of renting a 12-foot truck over the given distance.*

Distance	in miles	10	20	30	40	50
12-ft. truck	Cost (dollars)					

(2.10)

2. *Plot each point from table 2.10 on the grid provided and label clearly. Identify the y-intercept of the line that passes through these points.*

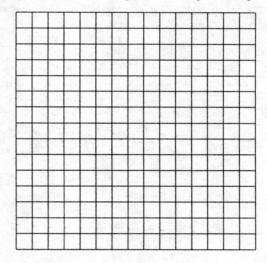

3. *Use the values in the table 2.10 to find the slope of the line in the graph above.*

4. *Determine the equation of this line. What do the variables x and y represent in this equation?*

5. *What is the relationship between the slope of the line and the per mile rate for truck rental charges?*

2.3.3   Problems

1. Recent statistics have shown the growing glut of Ph.D. degrees that have been awarded over the 20 years from 1975 to 1995. The following graph 2.11 represents the number of Ph.D. degrees (in thousands of degrees) from 1975 to 1995.

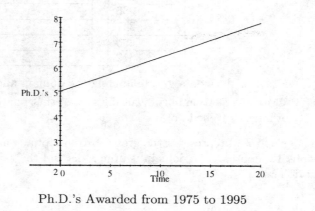

Ph.D.'s Awarded from 1975 to 1995   (2.11)

(a) Identify what the domain and range are. Explain.

(b) How many PhD's were awarded in 1975? in 1995? Explain. Write answers as ordered pairs.

(c) Find the rate at which Ph.D.'s were awarded over the 20 years from 1975 to 1995. Explain.

(d) Use your answers from ii. and iii. to write an equation that represents the line above. Explain.

(e) Use the equation in part d. to predict how many Ph.D.'s will be awarded in the year 2000. Explain.

2. A state's turnpike commission charges its users $0.04 per mile. There is a total of 333 miles of turnpike roadway. Answer each of the following questions.

(a) What is your toll between mile marker 226 and mile marker 312?

(b) What is your toll between mile marker 226 and mile marker 201?

(c) What is your toll between mile marker 236 and mile marker 298?

(d) Fill in table 2.12 (if you started traveling at mile-marker zero).

(2.12)

MILE MARKER		COST FROM MILE ZERO
201		
226		
236		
247		
298		
312		

(e) Plot this data on a graph labeling the domain and range clearly.

(f) Write an equation in two variables (x, y) that relates cost (y) in terms of miles travelled (x).

3. The graph 2.13 represents the growth in sales of a company over the previous 12 months. The domain is in months and the range is in thousands of dollars.

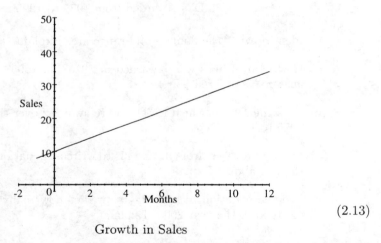

(2.13)

Growth in Sales

(a) How much in sales was the company making at the beginning of the year? at the end of the year? Write your answers as ordered pairs. Explain.

(b) At what rate are sales growing over the year? Label the rate clearly and show all calculations. Explain what the rate means.

(c) Write the equation of the line that represents modeling the company's sales over the last year.

(d) Use the model in part c above to determine projected sales 3 months from now. 8 months from now. Of these two projections which do you think is more accurate?

4. Statistics state that capital investment in cellular phone companies were rising over the year from January 1, 1997 ($t = 0$) to December 31, 1997 ($t = 12$). The graph 2.14 represents the amount in billions of dollars of capital investment in cellular phone companies over this 12 month period. The domain is in months and the range is in billions of dollars.

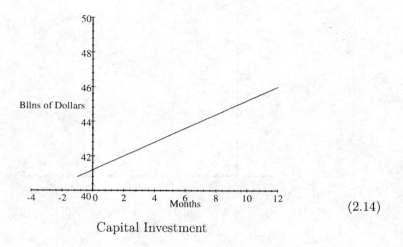

(2.14)

Capital Investment

(a) How many billions of dollars were invested in cellular phone companies at the beginning of the year? Write your answer as an ordered pair. Explain.

(b) How many billions of dollars were invested in cellular phone companies at the end of the year? Write your answer as an ordered pair. Explain.

(c) At what rate were investments changing over this year? Label the rate clearly and show all calculations.

(d) Write the equation of the line that represents modeling the capital investments over the last year.

(e) Use the analytic model in part 4 above to determine investments 3 months from December 31, 1997; 18 months from December 31, 1997. Of these two projections which do you think is more accurate? Why?

5. The statistics in problem 4 above also state that average monthly cellular phone bills declined over the same year. The graph 2.15 represents the average cellular phone bill over this 12 month period. The domain is in months and the range is in dollars.

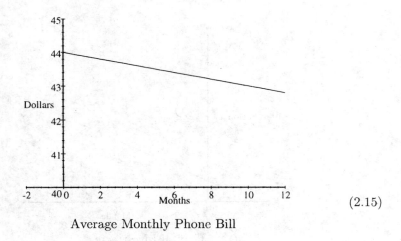

(2.15)

Average Monthly Phone Bill

(a) What was the average monthly bill at the beginning of the year? Write your answer as an ordered pair. Explain.

(b) What was the average monthly bill at the end of the year? Write your answer as an ordered pair. Explain.

(c) At what rate was the average monthly bill changing over this year? Label the rate clearly and show all calculations. Explain.

(d) Write the equation of the line that represents modeling the average monthly cellular phone bill over the last year.

(e) Use the analytic model in part 4 above to determine the projected average monthly bill 3 months from December 31, 1997; 18 months from December 31, 1997. Of the two projections you just figured out, which do you think is more accurate? Why?

6. The graph 2.16 represents the spread of all types of cancer in the state of Connecticut in the year 1982. The domain is in total population (in thousands of people) and the range is in thousands of people with cancer.

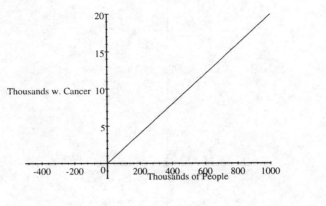

(2.16)

People with Cancer

a. Note that both the x-axis and y-axis above are written in terms of thousands of people. The y-intercept is (0,0). Explain what this means in terms of the number of cancer cases in this model. How many cases of cancer are there for 1,000,000 people (1000 on the graph)? Explain.

b. What is the rate of cancer cases per 1,000 people? Label the rate clearly and show all your calculations.

c. Write the linear model that represents the line in the cancer cases above.

d. Use the model you found in part c. to determine the number of expected cancer cases out of 2,000,000 people.

7. Rusty Automobile Company (RAC) has been marketing a new sports utility vehicle called the RRV. RAC has conducted a study to see if they can predict sales of the RRV over the next 8 years. The following graph 2.17 represents predicted sales of the RRV over the next 8 years. The range is in thousands of vehicles and the domain is in years.

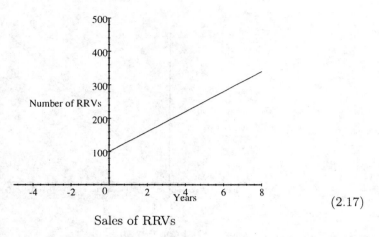

(2.17)

Sales of RRVs

(a) How many RRV's have been sold this year (that is, in the beginning)? in 8 years? Explain.

(b) At what rate does Rusty's expect the sales of the RRV to change? Explain.

(c) Determine an analytical model that supports the graph above? Explain.

(d) Use the model in part c to predict the sales of the RRV 10 years from now. Explain.

8. A state's sales tax on all non-essential items (non-clothes, non-food items) is 6%.

(a) Fill in the table entries for the total tax charged for the dollar amount indicated.

Purchase Value	in dollars	$10	$20	$30	$40	$50
Tax Charged	in dollars					

(b) Plot each point on a grid and label clearly. Identify the y-intercept of the line that passes through these points.

(c) Use the values in the table in part a to find the slope of the line in the graph above. Determine the equation of this line. What do the variables x and y represent in this equation?

(d) What is the relationship between the slope of the line and the per dollar rate for each purchase?

9. Much of the talk dealing with the "Information Superhighway" involves the speed with which information can be exchanged between systems. One of the fastest means of downloading and uploading digital bits of information currently is the cable modem (a system owed and operated by the cable companies). Another available system is a phone company technology called ADSL (asymmetrical digital subscriber line). Although the cable modem system is a little faster at delivering information than the ADSL system, ADSL is more secure and less expensive to the consumer (this represents in a small way the competition between and cable and phone companies to move into the new internet technology market), The ability to deliver data in kbps (kilobytes per second), both into and out of a system, for both the cable modem and the ADSL is listed in the chart.

System	Speed In (KBPS)	Speed Out (KBPS)	
Cable Modem	up to 2,000	up to 2000	(2.18)
ADSL	up to 1500	up to 512	

(a) Use the information in 2.18 that refers to **Cable Modem Speed In** to answer each of the following.

   i. Fill in the table entries for the time it takes to download information using this system.

Time	in seconds	10	20	30	40	50	60
Size of Data Set	in kilobites						

   ii. Plot each point on a grid and label clearly. Identify the y-intercept of the line that passes through these points.

   iii. Use the values in the table to find the slope of the line in the graph above. Determine the equation of this line. What do the variables t and y represent in this equation.

   iv. What is the relationship between the slope of the line and kbps?

(b) Use the information in 2.18 that refers to **Cable Modem Speed Out** to answer each of the following.

   i. Fill in the table entries for the time it takes to upload information using this system.

Time	in seconds	10	20	30	40	50	60
Size of Data Set	in kilobites						

   ii. Plot each point on a grid and label clearly. Identify the y-intercept of the line that passes through these points.

    iii. Use the values in the table to find the slope of the line in the graph above. Determine the equation of this line. What do the variables t and y represent in this equation.

    iv. What is the relationship between the slope of the line and kbps?

(c) Use the information in 2.18 that refers to **ADSL Speed In** to answer each of the following.

    i. Fill in the table entries for the time it takes to download information using this system.

Time	in seconds	10	20	30	40	50	60
Size of Data Set	in kilobites						

    ii. Plot each point on a grid and label clearly. Identify the y-intercept of the line that passes through these points.

    iii. Use the values in the table to find the slope of the line in the graph above. Determine the equation of this line. What do the variables t and y represent in this equation.

    iv. What is the relationship between the slope of the line and kbps?

(d) Use the information in 2.18 that refers to **ADSL Speed Out** to answer each of the following.

    i. Fill in the table entries for the time it takes to upload information using this system.

Time	in seconds	10	20	30	40	50	60
Size of Data Set	in kilobites						

    ii. Plot each point on a grid and label clearly. Identify the y-intercept of the line that passes through these points.

    iii. Use the values in the table to find the slope of the line in the graph above. Determine the equation of this line. What do the variables t and y represent in this equation.

    iv. What is the relationship between the slope of the line and kbps?

10. A company interested in getting into the long distance phone call business has launched a new phone card called CallWeb. To encourage people to use this new calling card the company is offering a "limited time offering" of 10 free minutes, after which customers will be charged $0.17 per minute when charging their calls to the card.

(a) Fill in the table below with the appropriate cost for number of minutes charged to the CallWeb card.

Total time on Card	in minutes	10	20	30	40	50	60
Charge to Card	in dollars						

(b) Plot the data from i. on a grid and label the axes and points clearly.

(c) Find an equation of the line in part b. using the data from a.

(d) Use the equation in c. to determine the cost of charging 2 hours to you CallWeb calling card.

(e) What is the y-intercept of this line? Does it appear on your graph? Why or why not? Explain.

11. Albert has recently retired. To fill his time he has decided to begin making and selling some hand-made crafts. These will be sold at a local craft show. A popular item among craft buyers is an earring holder. Albert has determined that the materials he needs to make an earring holder will cost $4.50 per holder. This particular earring holder requires a special paint brush and a special sealing paint. The paint brush costs $5 and the special paint costs $20 for a 5 gallon bucket. Albert has also decided to employ his grandsons to help produce his earring holders. Billy has agreed to paint all of the holders at a rate of $1.50 per earring holder. Joshua has agreed to package all of the earring holders for a salary of $15. Albert's only other cost is the $45 charge to have a table at the craft show. After some preliminary research Albert has decided to sell his earring holders for $10 each. Let x represent the number of earring holders that Albert will produce and sell. Answer each of the following questions.

(a) Determine Albert's cost function, $C(X)$ (in dollars), of producing x earring holders.

(b) Determine Albert's revenue function, $R(X)$ (in dollars) and Albert's profit function, $P(X)$ (in dollars), of selling x earring holders.

(c) Determine the number of earring holders that Albert must produce and sell in order to break even.

(d) Albert has determined that he needs $250 to purchase a diamond necklace he wants to give to his wife. How many earring holders should Albert produce and sell at this craft show?

12. In section 3.5 Albert was making animal puzzle sets and animal banks.

(a) Repeat the exercise using instead the fact that Albert will spend no more that $90 on material and no more than 150 hours.

(b) Determine Albert's maximum revenue if he sells each puzzle set for $9 and each bank for $15.

2.4   The Power of Linear Models: More Applications

Linear models are the simplest of all models to use. Every student knows the axiom that two distinct points determine a unique line. Thus, all we need to know about a problem are two distinct data points and we can develop a linear model representation of that problem. Of course, there are limitations to this approach. The world is not a linear one. In spite of this linear models are still quite useful. This section will explore some of the ways linear models are adapted to better represent certain problems.

2.4.1   Step Functions Models

Each of the examples that you worked out in sections 2.3.1 and 2.3.2 involved a linear model. That is, one that has a constant rate of change (slope) over the domain values. However, each of those examples (cellular phone charges and truck rentals) are not quite accurate in terms of how such companies charge their customers. Although the mathematics is correct there is still one further detail that needs to be added to each of the models so that they reflect the true rate at which these companies charge their customers.

Example 28 *Consider the example of the cellular phone company with a service that has a surcharge of $3.00 along with a $0.16 per minute charge. Actually, most phone services round the length of the calls made up to the nearest minute. For example, if you call someone and talk for 6 minutes and 19 seconds then the phone company charges you for a length of 7 minutes. Thus, the graph model of such a charge would look like the following:*

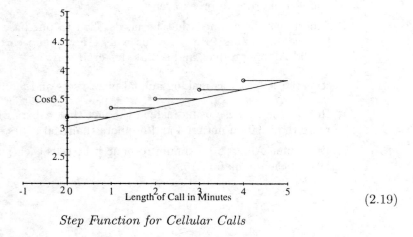

(2.19)

Step Function for Cellular Calls

In this graph you will notice a line along with a series of "steps." The line is not part of the model. It is placed here merely as a reference. The line represents the charge to a call if the company did not round up to the nearest minute as in the example in section 2.3.2. The "steps" indicate the charge to customers when the company does round up to the nearest minute. Take a moment to study the graph to see if it makes sense. The actual step function graph for this model would be

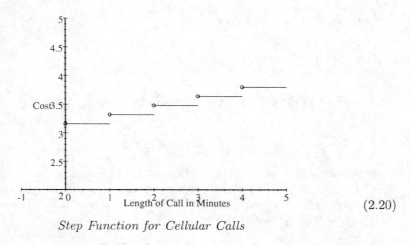

(2.20)

Step Function for Cellular Calls

Note that without the linear model as a reference the graph is harder to interpret.

Graphing a step function to model a case of rounding off is actually quite simple. To graph a step function

1. Develop a linear model first. This is done by, temporarily, ignoring any rounding off factor and treating the rest of the information in a linear fashion.

2. Once the line is graphed return to the rounding off factor and add this to the graph, rounding up or down as the situation indicates.

Such a graph is referred to as a **step function** (since it obviously resembles a series of steps) and can be written analytically by using one of two functions referred to as the **greatest integer function (or the round-down function)** and the **least integer function (or the round-up function)**.

Definition 29 *The **greatest integer function (or the round-down function)**, written $f_g(x) = \lfloor x \rfloor$, is defined to be the largest integer less than or equal to the value x.*

Definition 30 *The **least integer function (or the round-up function)**, written $f_l(x) = \lceil x \rceil$, is defined to be the smallest integer greater than or equal to the value x.*

Example 31 *Here are some examples:*

1. $f_g(3.7854) = \lfloor 3.7854 \rfloor = 3$

2. $f_l(3.7854) = \lceil 3.7854 \rceil = 4$

3. $f_g(-3.7854) = \lfloor -3.7854 \rfloor = -4$

4. $f_l(-3.7854) = \lceil -3.7854 \rceil = -3$

5. $f_g(6) = \lfloor 6 \rfloor = 6$

6. $f_l(6) = \lceil 6 \rceil = 6$

Therefore, in the case of the cellular phone company the new analytic model would be

$$f(x) = 0.16\lceil x \rceil + 3.00 \qquad (2.21)$$

where x represents the length of a phone call, $\lceil x \rceil$ represents rounding that call up to the nearest minute and $f(x)$ represents the charge to the customer. If the company chose to round down to the nearest minute instead of rounding up then the analytic model would be

$$f(x) = 0.16\lfloor x \rfloor + 3.00. \qquad (2.22)$$

For phone companies, rounding down to the nearest minute is not realistic. Examples of where rounding down would be realistic would be banks and investment accounts. When calculating the interest to their customers' investment accounts banks "truncate" off any fractional cents that are left. This is equivalent to the round-down function. Another example of applying the round-down function is the data collected by the United States Census Bureau.

Example 32 *Once every ten years the federal government is required by the constitution to determine the size of the United States' population. This accounting of the people is very important since these are the numbers that are used to make many important decisions about representation, federal funding, etc. The population continues to grow in the years following each census, however, any decisions the federal government makes in those intervening years are based upon the last census taken. The data from the last four censuses is listed in table ??. Note that the t-values for the domain are chosen so that graphing is more convenient.*

Year	1960	1970	1980	1990
t - value (domain)	0	10	20	30
Population (in millions)	179.3	203.3	226.5	248.7

As stated above we will first graph the linear model. Note that the census data is not quite linear. Actual data rarely is. However, it is "almost" linear and *so we will plot the line that **best** represents the data. This is referred to as **regression** - a topic we will revisit later in the text. On the graph 2.23 we have plotted the linear model (as a dotted line), the actual data (as diamond points) and the appropriate step function (soild lines). Note that the actual data is very nearly linear.*

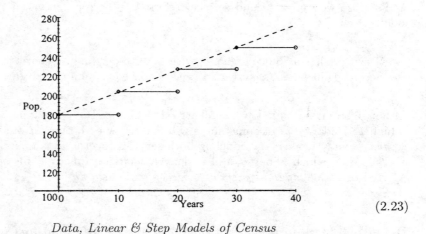

(2.23)

Data, Linear & Step Models of Census

In 2.24 we have plotted only the step function. the step function graphically represents how the government would interpret the population of the United States over the years from 1960 to 1999.

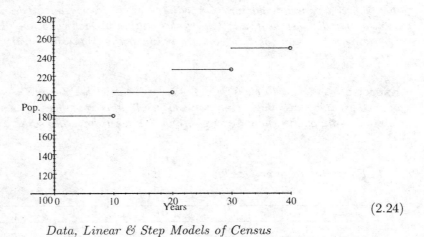

(2.24)

Data, Linear & Step Models of Census

Step functions occur in many situations in which rounding off is a part of the process. Several more examples of step functions are included in the problems at the end of this section.

2.4.2   Systems of Linear Models

When searching for a phone service, truck rental or some other professional service people are often interested in "comparison shopping." We wish to gather information on several services and decide which will be best for our needs. Recall the earlier problem where you are considering the cellular phone Service A. Suppose we find a second service, referred to here as Service B with the following charges.

Service A has a $3.00 surcharge with a $0.16 per minute charge.
Service B has a $4.00 surcharge with a $0.11 per minute charge.

Although Service A has a lower surcharge than Service B its "per minute" charge is higher. Under what circumstances will Service A be better for our use than Service B and vice-versa? Modeling both services together is an excellent way of comparing which is better and under what circumstances. This concept is referred to as *solving the system of models simultaneously.*

Definition 33 *A **system of models** or a **system of equations** is a set of equations*

$$y = f_1(x), y = f_2(x), ..., y = f_n(x) \qquad (2.25)$$

*that share a common domain. A point $(x_0, f(x_0))$ is said to be a common solution for the system if $(x_0, f(x_0))$ is a solution to each equation. The process of finding a common solution is called **solving the system simultaneously.***

Exercise 34 *As an in-class exercise we will model both services as a system of models and solve them simultaneously. In this instance the simultaneous solution is the point where Service A and Service B have the same cost.*

1. *Fill in the table entries for each service.*

Length of call (in min.)	5	10	15	20	25	30	35	40
Service A								
Service B								

2. *Plot each point on the grid provided and label clearly. Using the variables x and y write an equation that models the total charges for both Service A and B. Explain what x and y represent in the models.*

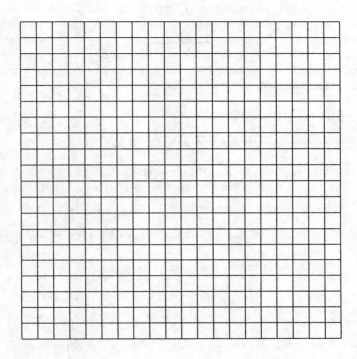

3. *Explain the significance of where the two lines intersect.*

4. *Using the graph above explain when Service A is better than Service B and vice versa.*

Clearly, where the Service A line and the Service B line intersect is significant. This is the point where Service B becomes more economical than Service A. Where the Service A line is "below" the Service B line is where Service A is more economical and where the Service B line is "below" the Service A line is where Service B is more economical. Therefore, it is to our advantage to know where these lines intersect. Once again, however, although the graph gives us a clear picture of what is happening it is inaccurate as far as the numerical values are concerned. We need an analytical way of simultaneously solving this system of models. Actually, there are several methods of solving systems of linear models simultaneously. We discuss two here.

Method One: The Substitution Method.

$$y = 0.16x + 3.00, y = 0.11x + 4.00$$

1. Solve one of the two models for one of the variables (in our example both models are already solved for y);

2. substitute this expression into the other equation

$$0.16x + 3.00 = 0.11x + 4.00;$$

3. Solve for the remaining variable

$$.05x = 1.00$$
$$x = 20 \, (\text{minutes});$$

4. Substitute this value back into either of the original equations and solve for the other variable

$$y = 0.16(20) + 3.00$$
$$= \$6.20$$

5. Write the solution as an ordered pair:

Service A and Service B generate the same cost for a 20 minute phone call $(20, 6.20)$.

Method Two. Linear Combination.

$$y = 0.16x + 3.00, y = 0.11x + 4.00$$

1. Line up one equation under the other, variable to variable, side to side.

$$y - 0.16x = 3.00$$
$$y - 0.11x = 4.00$$

2. Multiply both sides of both equations by appropriate quantities so that the coefficients of one of the variables are the same number but of opposite signs (in this example we choose x as the variable to demonstrate the technique, although it would have been easier in this case to choose the variable y.)

$$0.11(y - 0.16x) = 0.11(3.00)$$
$$-0.16(y - 0.11x) = -0.16(4.00)$$

$$0.11y - 0.0176x = 0.33$$
$$-0.16y + 0.0176x = -0.64$$

3. Now add the two resulting equations together to eliminate the one variable and solve for the remaining variable.

$$-0.05y = -0.31$$
$$y = \$6.20$$

4. Substitute this answer back into either of the original equations and solve for the remaining variable.

$$\$6.20 = 0.16x + 3.00$$
$$3.20 = 0.16x$$
$$x = 20 \text{ (minutes)}$$

5. Write the solution as an ordered pair:

Service A and Service B generate the same cost for a 20 minute phone call $(20, 6.20)$.

Several problems are included at the end of the chapter in the problems section. Try your hand at some of these before proceeding on.

2.4.3   Models of Linear Inequalities

In each of the sections 2.4.1 and 2.4.2 in this chapter we discussed some of the ways in which linear models can be applied to a number of different areas. However, in each case we were dealing with exact values in the sense that each domain element produced a distinct value in the range by plugging the domain value into the given function or equation. For example, there is an exact charge associated with each phone call of a specified length of time. There are instances, however, when exact values are not known. When we go the market we generally do not say "I will spend exactly \$54 today." It is more likely we will say something to the effect "I will spend no more than \$60 on groceries." To model this type of situation we rely on the mathematical concept of inequalities.

Definition 35 *A linear inequality in which y is the dependent variable (function) and x is the dependent variable is of the form*

$$ax + by \leq c \qquad\qquad (2.26)$$

or

$$ax + by \geq c. \qquad\qquad (2.27)$$

Note that we can simplify 2.26 and 2.27 to take either the form (not necessarily in this order)

$$y \leq mx + b \qquad\qquad (2.28)$$

or

$$y \geq mx + b. \qquad\qquad (2.29)$$

*in which case we say for 2.28 that y **is at most** mx + b and for 2.29 that y **is at least** mx + b.*

Here is an example.

Example 36 *Problem 3 from section 2.2.2 discussed the discovery of "lunar ice" by a scientific probe sent to the moon. Some of the information obtained from the probe was that there was enough water on the moon spread out over 25,000 square miles at the north and south poles to fill a lake 4 square miles and 35 feet deep. In that exercise you found that this implied that there was 6,183,260 cubic feet per square mile or approximately 6.2 million cubic feet of water per square mile. If we let x be the number of square miles (the domain) and f(x) be millions of cubic feet of water, then an analytic model for this problem is*

$$f(x) = 6.2x. \qquad\qquad (2.30)$$

*(Note that there is no water in 0 square miles so the y-intercept is (0,0). This information is in fact an **upper estimate**. This means there is **at most** 6.2 million cubic feet of water per square mile. That means that the equation 2.30 is not completely correct since the equation implies that there is **exactly** 6.2 million cubic feet of water per square mile. Therefore, a much better representation of the given information in this case is the following inequality model*

$$f(x) \leq 6.2x. \qquad\qquad (2.31)$$

Suppose we are interested in knowing how much water would be contained in 1000 square miles. Then we would have

$$f(1000) \leq 6.2(1000) = 62,000,000,000 \ cubic \ feet \ of \ water$$

*and we would say that "there is **at most** 62,000,000,000 cubic feet of water in 1000 square miles located at the poles of the moon.*

We could also easily construct a geometric model of this information by graphing the inequality $f(x) \leq 6.2x$ in the following way

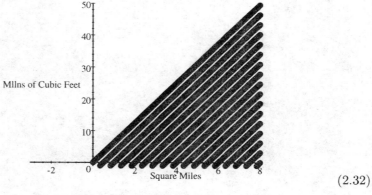

(2.32)

Inequality for Moon Ice

The shaded region above represents all points in the plane that satisfy the inequality 2.31. To graph the region 2.32 is quite simple.

1. First, as in the case of step functions, graph the line associated with the equation 2.30; this line is the boundary of the region.

2. The line 2.30 divides the plane into two regions, one "above" the line, one "below" the line. One of these regions represents the "at most" case ($f(x) \leq 6.2x$) while the other region represents the "at least" case ($f(x) \geq 6.2x$). Do not make the mistake of thinking that the "upper" region is automatically the "at least" case or that the "lower" region is the "at most" case.

3. To determine which of the two regions is the correct one for the model given requires a simple calculation. There are many ways to do this. This most straight-forward way is to simply pick a point from one of the two regions and substitute it into the inequality. If the resulting statement is true then the region from which the point was chosen is the correct region. If the resulting statement is false then the region from which the point was chosen is the incorrect region and the other region is the correct one. For example, the point $(2,0)$ is not on the line and is in the "lower" region. Substituting this point into the inequality 2.31 produces $0 \leq 6.2(2)$, which is true. On the other hand, the point $(2,20)$ is in the "upper" region. Substituting this point into the inequality 2.31 produces $20 \leq 6.2(2)$, which is false. Hence, the "lower region" is the one to be shaded.

The analytical model 2.31 also gives a great deal of information about the problem.

Exercise 37 *Use model 2.31 to answer each of the following.*

1. *How much water could scientists expect to find within a 4 square mile area?*

2. *In planning for a lunar station, scientists calculated that occupants of the station would require about 10,000 cubic feet of water per day. If the station is to be operational for one year, how large of an area on the moon's surface will be needed to support the station?*

2.4.4   Linear Programming Models

In this section we look at yet another way of applying linear models to problems. Linear programming is an application that combines several of the concepts discussed in the previous sections. Applications include such fields as business and operations research, any time we wish to optimize our options. We will introduce this concept by way of an example.

Example 38 *Albert has recently retired. To fill his time he has decided to begin making and selling some hand-made crafts. These will be sold at a local craft* show. He has decided to produce animal puzzle sets and animal banks. The animal puzzle sets require six hours of assembly time while the animal banks require only two. His material costs are $2.50 per animal bank and $1.50 per animal puzzle set. Albert wants to spend no more than 120 hours of preparation time for the show and no more than $75 for materials.

1. First notice that Albert is interested in building two types of objects; animal puzzle sets and animal banks, each of which requires a certain amount of time and resources. In a sense they depend on each other since the more of one that is built then the less resources there are to build the other. Since Albert has set aside 120 hours and $75 to complete his project the time and the money must be divided between the puzzle sets and the banks. We can organize the information in a chart to assist in the solution. Let x be the number of puzzle sets that he makes and let y be the number of banks that he makes. Then

	Amount Spent on Puzzle Sets	Amount Spent on Banls
120 Total Hours	$6x$	$2y$
$75 Total	$1.50x$	$2.50y$

$$(2.33)$$

Using the information in chart 2.33 we can write a system of equations that represents this situation where Albert spends exactly 120 hours of preparation time and exactly $75 for materials.

$$6x + 2y = 120 \ (total \ hours), \ 1.50x + 2.50y = 75 \ (total \ time) \quad (2.34)$$

2. Use one of the available methods to solve that system of equations for the number of animal puzzles and animal banks Albert must sell. This answer represents how many puzzle sets and banks Albert must sell to use up all of his time and money resources. However, it does not tell us if this is the best answer to optimize his profits. To find out how to do this read on.

3. Of course, Albert does not need to use up all of his resources. His can spend at most 120 hours and at most $75 working on his retirement project.

Write a system of linear inequalities that represent Albert spending at most 120 hours and at most $75 on his project.

4. Sketch a graph of the system of linear inequlaities found in part 3 above.

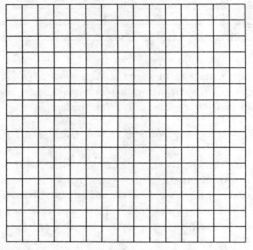

The shaded region that is common to both inequalities is the "good" region. That is, any point from this region represents a certain number of puzzle sets and banks that Albert can make and still remain within his constraints for time and money.

5. Next, how can Albert maximize his revenue based upon these constraints? His profit will depend on at what price he can sell each puzzle set and bank. For now let's assume that, for whatever reasons, he has determined that he must have at least eight of the banks and at least of the puzzles at the show in order to pay for the registration fee. Each animal bank sells for $12 and each puzzle set sells for $8. His revenue then is determined by the partial expression

$$8x + 12y.$$ (2.35)

We will use expression 2.35 to determine how much revenue Albert can make within his limitations. For example, is it possible for Albert to earn $250 in revenue within his limits? To answer this add the "revenue" line

$$8x + 12y = 250 \qquad (2.36)$$

to the graph above. If the line crosses over any part of the "good" region then this implies that there exist points (x, y) that satisfy both the revenue 2.36 and the restrictions of 2.34. Thus, we simply continue to increase the revenue in 2.36 until the line "exits" the region. The last line with a point within the region is the maximum revenue. Fill in the chart to determine the maximum revenue.

Revenue	$250	$280	$300	$320	$340	$360	$380	$400
Good	yes							

2.4.5   Problems

1. Many truck rental companies as well round up to the nearest mile or the nearest ten miles, etc. Use the information from the truck rental example from section 2.3.2 to work on the following exercise. Recall that a local truck rental franchise that you own rents 12-foot trucks at a flat $40.00 surcharge up front plus $0.31 per mile. Suppose also, that you decide to round up the charges to the nearest 10 miles.

    (a) Re-plot the line for the original data on a grid and label clearly.

    (b) Next, plot a step function that represents the fact that you will be rounding each charge up to the nearest 10 miles.

    (c) Determine the analytic (step) model that represents rounding up the distance to the nearest ten miles.

2. Problem 8 in section 2.3.3 discusses a 6% sales tax for a certain state. For this problem graph a step function that represents charging a 6% sales tax on the nearest whole dollar amount greater than the purchase price (rounding up the purchase price to the nearest dollar amount). Give the analytic model that represents this.

3. Problem 9 in section 2.3.3 discusses various speeds for different types of modems. The speed given is an upper estimate. Due to the number of users currently on line, as well as other factors, the actual speed may be less. For each part of problem 9 graph a step function that represents the upper limit for each minute of use. Give the analytic model that represents this.

4. City parking garages usually charge by the hour. They also round up to the nearest hour. For example, if you were to park your car for 3 hours and 17 minutes, you would be charged for the full 4 hours.

    (a) Determine a linear model that represents the case where a parking garage charges $4 per hour for parking.

    (b) Graph the linear model on a grid. Label the axes clearly.

    (c) Next, plot a step function that represents the fact that the garage will be rounding each charge up to the nearest hour.

    (d) Determine the analytic (step) model that represents rounding up to the nearest hour.

5. You recently acquired a local truck rental franchise that rents 12-foot and 20-foot trucks (see section 2.3.2). After assessing the dynamics of the market and your possible costs you decide that a fair rental fee for the 12 ft trucks is to charge customers a flat $40.00 surcharge up front plus $0.31 per mile. You also decide that a fair rental fee for the 20 ft trucks is to charge a flat $48.00 surcharge plus $0.16 per mile. Use this information to answer each of the following.

(a) Fill in the table entries for each truck size.

Distance	in miles	10	20	30	40	50	60
12-ft. truck	Cost (dollars)						
20-ft. truck	Cost (dollars)						

(b) Plot each point on a grid and label clearly. Using the variables x and y write an equation that models the total charges for both the 12-foot truck and the 20-foot truck. Explain what x and y represent in the models.

(c) Explain the significance of where the two lines intersect.

(d) Using the graph above explain when it is better to rent the 12-foot truck over the 20-foot truck and vice versa.

(e) Use either the substitution method or the linear combination method (or another method you may have learned elsewhere) to analytically solve the equations simultaneously to determine when the cost of renting the 12-foot truck and the 20-foot truck will be the same.

6. During an eight hour day at the stock exchange, stock A (that of an oil company) started the day at $61 and dropped steadily at an average rate of $0.40 per hour while stock B (that of a high tech company) started the day at $42 and rose steadily at an average rate of $0.30 per hour. Answer each of the following questions.

(a) Fill in the table entries for each stock price.

Time	in hours	1	2	3	4	5	6
Stock A	Price (in dollars)						
Stock B	Price (in dollars)						

(b) Plot each point on a grid and label clearly. Using the variables t and y write an equation that models the price of each stock throughout the day. Explain what t and y represent in the models.

(c) Explain the significance of where the two lines intersect.

(d) Using the graph above explain at what point in the day Stock A is priced higher than stock B and vice versa..

(e) Use either the substitution method or the linear combination method (or another method you may have learned elsewhere) to analytically solve the equations simultaneously to determine when the price is the same for both stocks.

2.5   Summary

The linear model is the first of the mathematical models we will study. It is also the easiest to develop. All that is required is two data points. As a result the linear model can be represented by geometric, analytical or numerical means.

In spite of its simplicity there are many applications for which it is perfectly suited. A good rule of thumb when modeling a problem is to keep the model as simple as possible. Unless accuracy dictates a more complex approach the model chosen should be as simple as possible. As a result the linear model is used quite often. Some of the ways in which it is used can be seen with round-off functions, systems of linear models and linear inequalities.

We will study other types of models in this text, however, this will not diminish the usefulness of the linear model.

Chapter 3

Models of Ratio and Proportion

3.1 Introduction

One of the most common and useful of concepts in mathematics is the concept of *comparison*. Comparisons between two quantities are used in order to acquire necessary information about a situation or problem. For example, many people hire financial advisors to help manage their personal assets. Financial advisors offer advice on how to invest, in what to invest, when to invest, etc. Often a detailed analysis of a person's financial position is used to provide the advisor with enough information to make intelligent and informed suggestions. Table 3.1 below is an example of two customers' financial summaries that the advisor would use to help give the appropriate financial or investment advice.

Categories	First Investor	Second Investor	
Assets	$93,568	$237,542	
Liabilities	1,700	199,300	
Net Worth	**$91,868**	**$38,242**	(3.1)
Annual Income	$62,114	$83,245	
Annual Expenses	53,778	79,332	
Discretionary Income	**$8,336**	**$3,913**	

Note that the second investor has more than twice the assets of the first investor, however, the first investor's liability is much smaller than the second investor's liability. The financial advisor would *compare* liabilities to assets in order to understand which customer is in the better financial shape. That is,

$$\frac{\text{First Investor's Liabilities}}{\text{First Investor's Assets}} = \frac{\$1700}{\$93568} = \frac{25}{1376} = 0.018169 \quad (3.2)$$

$$\frac{\text{Second Investor's Liabilities}}{\text{Second Investor's Assets}} = \frac{\$199300}{\$237542} = \frac{99\,650}{118\,771} = 0.83901$$

When these comparisons in 3.2 are converted to a common decimal it is easy to see that the first investor's assets far outweigh their liabilities.

Such comparisons in mathematics are called *ratios*.

Definition 39 *A **ratio** is a comparison between two quantities. Two ratios, $\frac{a}{b}$ and $\frac{c}{d}$, are said to be **in proportion** if they are equal, $\frac{a}{b} = \frac{c}{d}$.*

The concept of the slope of a line from Chapter 2 is a ratio. Slope of a line compares the change in the vertical quantity with respect to the change in the horizontal quantity. Percentages are another common use of ratios.

We will return to table 3.2 in a later chapter where we will discuss financial models. In this chapter we will demonstrate several other practical ways in which the common ratio is useful.

3.2   Population Characteristics

3.2.1   Tagging: Estimating Population Size

We often use ratios (fractions) and proportions to find an unknown quantity by comparing it to a proportionally equivalent, known quantity. A good example of this is estimating the size of a "moving" population. An environmental concern is that of endangered species. Those species endangered by over hunting, over harvesting or destroyed habitat must be monitored very carefully in order to determine how many of the species are left in "the wild." This is accomplished by the technique of **tagging**.

Tagging: A certain number of the species are first captured. They are then tagged and released back into the wild. The total number of tagged species will be represented by T. The tagged specimens are given some time to re-mingle with the general population. It is assumed that the species randomly mingle with the rest of the population. Observers are then sent out to various locations with the job of counting each of the following: the total number of tagged species seen (T_S) and the total number of species seen altogether (N_S). If N is the total number of species in the wild and if we assume that the species members mingle among themselves randomly, then we can estimate the value of N by using the proportion

$$\frac{T_S}{T} = \frac{N_S}{N} \qquad (3.3)$$

The only unknown quantity in the proportion 3.3 is N. Thus, we simply substitute into 3.3 the values that were found for each of T_S, N_S, and T and solve for N.

One such example that is in the news from time to time is when commercial species such as lobster and salmon are over-harvested. As a result these commercially rewarding species must be monitored so that they are not harvested to extinction. Here is how tagging works.

Example 40 *Suppose that we are interested in estimating the number of an endangered fish in a small lake. First, scientists capture several fish which are then, counted, tagged (the total number of tagged fish will be represented by T) and then released back into the lake. It is assumed that after the tagged fish are released into the lake they randomly "mingle" among the fish population. Next, teams of "counters" are sent to several randomly selected locations around the lake to capture and count fish in the following manner: Each team counts the total number of fish caught (N_S) and the number of tagged fish caught (T_S). When the teams return with their fish counts and the amounts are totaled, the fish population can be estimated with this information by using the following proportion formula:*

$$\frac{\text{Number of tagged fish }(T_S)\text{ counted by teams}}{\text{Total number of tagged fish }(T)\text{ in lake}} = \frac{\text{Total number of fish }(N_S)\text{ counted by teams}}{\text{Total number of fish }(N)\text{ in lake}}$$

Since three of these values are obtained through the counting procedures described above, the number of fish in the lake can be estimated by plugging in the appropriate values and solving for N. In your own words answer each of the following.

1. *Note that this technique only <u>estimates</u> the number of fish in the lake. How can we be sure that our method is producing a good estimate? Discuss in class possible ways in which we could increase our confidence that the estimate we obtain is a "good" one.*

2. *Can you describe another application of this technique? Explain.*

3. *Describe how this technique might be adapted to estimate the number of students that are enrolled in your school.*

This technique can also be demonstrated in class with the following exercise.

Exercise 41 *The class instructor should bring to class a large number of objects (a jar of pennies, a box of bottle caps, paper clips, etc.). Have the students estimate the number objects by the tagging technique.*

1. *Describe below a way by which the objects could be tagged.*

2. *Outline the technique (include calculations) used to estimate the number of objects.*

 3. Explain how to increase confidence in the estimate found in part 2 above.

3.2.2 Analyzing Population Behavior

In Chapter 2 we discussed the concept of slope of a line. The slope of a line was described as the rate at which the range value changed with respect to how the domain value changed. That is, the slope is the *ratio between the change in the range and the change in the domain*

$$slope = \frac{\text{change in range value}}{\text{change in domain value}}.$$

In the case where the y-intercept of the line is the origin, the range represents the size of a population and the domain represents time, then the slope represents the rate at which the population is changing over time. Knowing the rate at which a population changes over time is important since it allows us to predict the size of the population at some time in the future. The word *population* is used here to simply mean a set of objects. A population can be people, insects, money invested in an account, the capacity of containers; any group of objects whose behavior we want to study. For example, being able to predict how large a population will be means that we can plan for future needs. This may mean analyzing how best to invest an appropriate amount of money in a retirement plan now in order to assure a comfortable retirement income.

Another example would be quality control. Companies that offer goods or services for sale often test the reliability and consistency of their product. By monitoring the product "population" customers can be assured of a consistently high standard of product quality. Work through the following exercise in class.

Exercise 42 *(Quality Control) TAE Electric Co. publishes a statistic that their light bulbs will last at least 300 hours. The quality control at TAEEco attempts to limit the number of defective bulbs (that is, those that fail in less than 300 hours) to less than 15 in 10,000.*

 1. What is the expected defect rate as a percentage?

2. *Your school just purchased 1500 light bulbs which are in service approximately 1.5 hours per day, 200 days per year. How many of these light bulbs would you expect to fail by the end of the school year?*

3. *By the end of the school year 18 bulbs have failed. Do you have reason to complain to TAEEco?*

4. *What other external factors could affect the performance of the light bulbs at your school? That is, what conditions might exist on-sight that will affect the light bulbs' performance that might not exist in TAEEco's labs? Does this change your answer to part 3? Explain why or why not.*

Most products are tested under laboratory conditions. Although the company may try to simulate actual conditions as much as possible, they cannot anticipate every condition that may actually occur. The question that you must answer is whether or not this justifies the failure of 18 light bulbs out of the 1500 originally purchased.

This exercise also asked that some of your ratios be represented as percentages. Clearly, a percentage is a special type of ratio. *Percent* is a Latin-based word meaning "out of one hundred." We use percents to describe how much of a population exhibits a certain characteristic. This is done by stating how many "*out of every one hundred*" of the population have that characteristic. The percent is found by the formula

$$\text{Percent} = \frac{\text{the number of the population with the characteristic}}{\text{the total number of the population}} \times 100.$$

What confuses many students is the fact that the percent notation is just that, a notation. The notation 73% is simply used for notational convenience. The notation 73% would never be used when solving a mathematical equation or simplifying a mathematical expression. The numerical equivalent of 73% that is used when manipulating mathematical expressions is the decimal quantity .73 or the rational quantity $\frac{73}{100}$ since the notation 73% means "73 out of every 100." We will return to percent and its applications later in the text.

3.2.3   Problems

1. Using the technique of tagging to...

    (a) estimate the number of students that are enrolled in your school.

    (b) estimate the number of students that are left-handed in your school.

    (c) estimate the number of students that are wearing something with the color red in it.

    (d) estimate the number of faculty members at your school.

2. Statistics have shown that each year we can expect to see about 20 new cases of cancer per 1000 people. Use this information to answer each of the following questions.

    (a) Represent this statistic as a percentage.

    (b) About how many cases of cancer can be expected to occur this year in a city of 500,000 people?

    (c) If it is also known that 6 of these 20 new cases will prove to be fatal this year, how many deaths in this city can be expected to be attributed to cancer this year?

    (d) About how many cases of cancer can be expected to occur this year in your area?

    (e) If it is also known that 6 of these 20 new cases will prove to be fatal this year, how many deaths in your area can be expected to be attributed to cancer this year?

    (f) Most statistics are gathered through random population sampling and observation techniques. What external factors could affect our interpretation of these statistics?

3. There is a great deal of concern over global warming trends and the depletion of the earth's ozone layer. As a result scientists have begun to monitor the level of certain greenhouse gases in the atmosphere. Through observation techniques scientists have been able to determine the current amount of certain greenhouse gases in the atmosphere. The table below gives the amount in parts per million (ppm) of some of the more common greenhouse gases. Use this information to answer each of the questions below.

Greenhouse Gases	Amount in Atmosphere in ppm
Methane	1.72

    (a) Represent each statistic as a percentage.

    (b) If we assume that the gas is distributed uniformly through the atmosphere determine the amount of gas contained in a one cubic mile region.

(c) If you inhale ....amount of air with each breath and you take .....breaths per minute, how much of the gas will you inhale in 12 hours?

(d) Most statistics are gathered through random population sampling and observation techniques. What external factors could affect our interpretation of these statistics?

4. States often pas laws that will insure the environmental safety of children. For example, a state may set water purity standards that require household water to be 99.9% lead-free. We discover a household where there is a 1 part per 100 lead concentration.

(a) Is this level of lead concentration acceptable? Explain your answer and support it with the appropriate mathematical computations.

(b) A water filtration system that attaches directly to a faucet is available in local stores. Its manufacturer claims that it will remove 98% of all remaining lead found in water. Is this filtration system acceptable to bring the lead content in the household to within state-mandated guidelines? Explain your answer and support it with the appropriate mathematical computations.

(c) Since the water filtration system was most likely tested under laboratory conditions, what external factors could affect our interpretation of these statistics?

5. Optical Character Recognition (OCR) packages are a type of software package that scans and analyzes printed documents and then converts the documents to electronic text. An OCR is said to be **90% effective** if there are no more than 10 incorrectly translated characters for every 100 characters the package attempts to translate. The more effective a package is the more expensive it is to purchase. Suppose that there are three OCR packages that fall within the desired budget.

OCR package A is 90% effective,

OCR package B is 99% effective, and

OCR package C is 99.99% effective

(a) If a standard 8×11 page contains 80 characters per line and 66 lines, how many errors per page do you expect from each of these OCR packages?

(b) Obviously, OCR packages do not identify where the translation errors have occurred in the document. Locating translation errors is the job of a proofreader. Of course, this takes time. As a proofreader of documents, discuss what is an acceptable number of corrections per page that you are willing to make.

(c) Use your answer in part b to determine if any of the OCR packages above will suffice for your proofreading needs.

(d) A 100% effective OCR package is obviously preferable but unattainable in practice. The more effective an OCR package is, the fewer translation errors exist. However, the fewer errors there are, then the harder it is to find them. If OCR package D is 99.999% effective, is it more desirable to have than package C?

(e) Repeat part d for OCR package E which is 99.9999% effective.

6. Statistics from the NHSTA indicate that over a ten year period there were 2.6 million automobile accidents in the United States that involved airbag deployment. They also report that 66 of those 2.6 million accidents resulted in a child's death.

(a) Represent the child fatality rate as a percentage?

(b) A US city reports that 9800 accidents involving airbag deployment were recorded by their city in 1998. How many child deaths should city officials expect as a result?

(c) Another city reports that there were 8 child fatalities as a result of accidents involving airbags. Estimate how many accidents involving airbag deployment occurred in that city.

(d) Using a local source (such as your library, the internet, etc.) conduct a search for statistics on the number of accidents involving airbags in your area. Use this information to estimate the number of child fatalities that could be expected.

7. In order to reestablish the environmental balance in a region ecological organizations attempt to reintroduce a native species into the region's environment. This is done by various techniques which include raising the species in captivity and then releasing them into the environment or transporting them from one area to the desired area. Such methods of repopulating a species often entail risks. Animals raised in captivity do not always adapt successfully to being in the wild. Animals transported from one environment to another do not always remain there. A local ecological group has conducted a study and found that a particular species reintroduced into an environment has a 1 in 5 chance of surviving in that environment. That is, for every five animals reintroduced into the environment, one is expected to survive.

(a) If 47 of the species are introduced into the environment, how many can be expected to survive?

(b) If at least 36 of the species are needed to provide an adequate gene pool for the species to reproduce on its own in the wild, how many animals should the environmentalists release into the wild?

(c) If 127 animals were released into the environment and 10 were observed to have remained and survived, give some explanation as to why this is the case.

3.3   Direct Comparisons

Ratio and proportion concepts also play an important role in comparing other types of quantities. We present here three additional examples of how ratios and proportions are use to gather some information about a particular situation.

3.3.1   Measuring Diversity

One way in which botanists measure the "health" of a forest is by determining how diverse the tree life in the forest is. That is, they compare the number of different species and how they are distributed through the forest. There are many ways in which this can be accomplished. The following exercise demonstrates one such way.

Exercise 43 *The diagram 3.4 below is a depiction of a small forest. The squares represent oak trees, the diamonds represent maple trees and the small circles represent birch trees. This exercise describes one way to measure the diversity of tree life in a forest.*

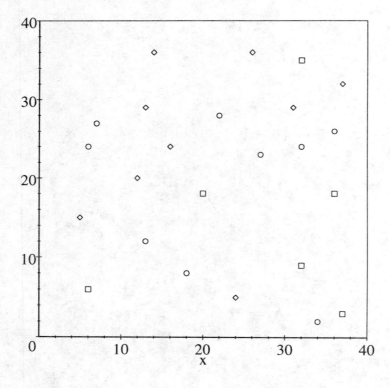

$$(3.4)$$

1. *Use a ruler to subdivide the forest into 16 equal subregions. When you do this you will notice that some trees may fall partly in more than one section. Discuss a rule by which you will define a tree to be "contained in" a section.*

2. *The **relative density of a tree in a forest** is the ratio between the number of sections that contain that tree and the total number of sections. Compute the relative density of each type of tree.*

3. *The **relative frequency of a tree in a forest** is the ratio between the number of trees of that type in the forest and the total number of trees in the forest. Compute the relative frequency of each type of tree.*

4. *Use the information that you found in parts 2 and 3 above to describe the occurrence of oak trees, maple trees and birch trees in our forest.*

5. *Use the information that you found in parts 2 and 3 above to calculate a single quantity whose value will identify the "diversity" of tree-life in the forest.*

3.3.2 Measuring Porosity

Another application of ratio and proportion arises from the need to know how quickly water drains through different soil types. Among other things, this information is used by highway construction crews to determine what kind of bedrock is needed to support highway drainage as well as by geologists to determine the possibility of flooding in a particular region. Although the calculations for these applications would be in three dimensions the class exercise presented here is in two dimensions. The three dimensional case is presented as an exercise at the end of the section.

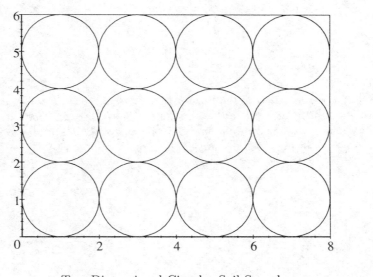

(3.5)

Two-Dimensional Circular Soil Sample

Exercise 44 *Porosity of soil is a measure of the "air space" between the soil particles. In geology soils are classified into two primary types: sand and clay. Sand consists of granular particles that retain none of the water that passes through it. Clay consists of extremely small particles that retain some of the water that passes through it. When measuring the porosity of soil we wish to determine the proportion of air space to soil particles. One way to accomplish this is by comparing the amount of air space to the total amount of air and soil. As an in-class exercise consider the two dimensional diagram 3.5 of "sand particles."*

   1. *Using diagram 3.5 compute the total amount of "soil" and "air." Explain the rational behind your model.*

2. *Using diagram 3.5 compute the total amount of "air." Justify your procedure.*

3. *The **discharge ratio** is the ratio between the amount of air space to the total amount of air and soil. Compute the discharge ratio for the two dimensional soil sample in diagram 3.5. What are the minimum and maximum values of the discharge ratio? Explain.*

4. *The **porosity** is the ratio between the amount of air space and the amount of soil. Compute the porosity for the two dimensional soil sample in diagram 3.5. What are the minimum and maximum values of porosity? Explain.*

5. *By using more than one measurement we can often gain additional insight to the problem being modeled. Compare the advantages and disadvantages to using discharge ratio and porosity.*

In practice it is impractical to directly measure the porosity or discharge ratio of the soil as the previous exercise implies. However, from some practical observations a method to approximate the soil's porosity can be devised. We will investigate this process in the following exercise.

Exercise 45 *Consider the measurements of discharge ratio and porosity discussed above.*

1. *What can you say about the relationship of the two measurements? For example, when the discharge ratio increases, what happens to porosity?*

2. *If water flows through a soil sample from top to bottom, under what circumstances would you expect the water to flow more quickly? Less quickly? That is, for what values of discharge velocity and porosity will the water flow more or less quickly?*

3. *The flow of a liquid through a granular substance is called* **percolation**. *The rate of percolation is a measurement of the "quickness" of the flow of the liquid through the substance (e.g. water through soil). How do you think percolation rate is related to discharge ratio? To porosity? Explain your answer.*

4. *Using your observations, devise a method to measure soil porosity. Explain how the procedure yields a measurement and how the behavior of that measurement can be interpreted to compare the porosity of various soil samples.*

There are many applications where percolation of a liquid through a granular substance is used to complete a process. In most cases, the size of the grains in the granular substances is used to control the percolation rate of the liquid. In applications where only a measurement of porosity is required, usually a comparative measurement of the percolation rate is made. Several problems appear at the end of this section that demonstrate this concept.

3.3.3 Measuring Height

The versatility of ratio and proportion is evident in this next example. When a NASA probe lands on another planet scientists are interested in determining the height of the surrounding hillside. The techniques they use are similar to the techniques used for determining the height of tall buildings, trees or distant objects. One technique is demonstrated in the following exercise.

Exercise 46 *Length-measuring tools such as meter sticks and tape measures are used to measure distance or the length of an object if the distance is not very far or if the object in question is small enough. However, it is extremely impractical to measure a tall skyscraper by placing a meter stick end over end along the side of the building. A more practical method to measure the building's height must be found. There are many methods for measuring great distances or great heights. These methods usually involve an indirect mathematical approach. One such method is outlined in this exercise. Although meter sticks cannot be used directly to measure the height of a building, they can be used indirectly to do so. Students should work in pairs for this exercise.*

1. *Choose one member of the pair and measure their height using the meter stick. Record the results below. Describe the accuracy of the measurement.*

2. *Go outside the building. The student whose height was measured (Student One) should stand up against the building. The other student (Student Two), holding the meter stick up toward the building walks away from the building until the building "appears" to be exactly one meter in height looking through the meter stick. At that point Student Two also measures the height of Student One as it appears through the meter stick.*

3. *The true height of the building can now be estimated by the following proportion.*

$$\frac{\textit{Actual height of Student One}}{\textit{Apparent height of Student One through meter stick}} = \frac{\textit{Actual height of Building}}{\textit{Apparent height of Building (1 meter}}$$

Substitute your measurements into this proportion and estimate the height of the building.

4. *This procedure certainly will generate some error in the measurements. Describe some of the ways measurement error can occur and what can be done to improve the accuracy of the measurements.*

This technique of measuring the height of a building is based upon the geometric principle of **similar triangles**.

Definition 47 *Two triangles, $\triangle ABC$ and $\triangle DEF$, are said to be **similar**, $\triangle ABC \sim \triangle DEF$, if their corresponding sides are in proportion. That is, if $\frac{AB}{DE} = \frac{BC}{EF} = \frac{AC}{DF}$.*

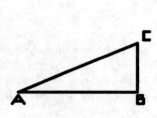

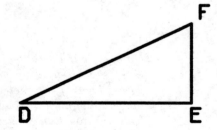

Triangle ABC

Triangle DEF

(3.6)

Informally, we say that $\triangle ABC$ and $\triangle DEF$ are the same "shape" but are of different "sizes" (see diagram 3.6). An equivalent definition is the following.

Definition 48 *Two triangles, $\triangle ABC$ and $\triangle DEF$, are said to be similar if $\angle A = \angle D, .\angle B = \angle E$ and $\angle C = \angle F$.*

Study diagram 3.6 for awhile to become convinced that these two definitions say the same thing. In fact, both definitions will be used in setting up a special case of ratio and proportion which will prove very useful. This special case is based upon right triangles. Consider the case when $\triangle ABC$ and $\triangle DEF$ are right triangles (with the right angles at vertex A and vertex D, respectively, as in diagram 3.6). The study of the properties of right triangles is called **trigonometry**. The **trigonometric ratios** (or **functions**) are defined by the ratios of the sides of the triangle.

Trigonometry is entirely based upon the relationship between the ratios of the sides of a triangle and its corresponding angles. The two definitions above on similar triangles imply that *any time two triangles have the same angle measures then the sides are in proportion*. This means that as long as we know the angles of a triangle then we automatically can identify what the ratio between the sides of that triangle should be. In order to keep things simple we organize the ratios of the sides of the triangle according to an angle of reference. For example, consider $\triangle ABC$ and $\triangle DEF$ in diagram 3.6 above. Since the right angles are located at vertex A and vertex D we will use another angle as the reference angle, say the angles at vertex C and vertex F. Angle C and angle F are corresponding angles. That is, they appear in the relative position in each triangle. The side opposite angle C in $\triangle ABC$ is $\overline{AB}$. The side opposite angle F in $\triangle DEF$ is $\overline{DE}$. The side adjacent to angle C in $\triangle ABC$ is $\overline{AC}$. The side opposite angle C in $\triangle DEF$ is $\overline{DF}$. The hypotenuse in $\triangle ABC$ is $\overline{AC}$. The hypotenuse in $\triangle DEF$ is $\overline{EF}$. Each triangle produces three possible ratios between the sides

$$\frac{AB}{AC}, \ \frac{BC}{AC} \ \text{and} \ \frac{AB}{BC} \tag{3.7}$$

in $\triangle ABC$ and

$$\frac{DE}{DF}, \ \frac{EF}{DF} \ \text{and} \ \frac{DE}{EF} \tag{3.8}$$

in $\triangle DEF$.(Actually, there are six possible ratios in each triangle, the three listed in 3.7 and in 3.8 and their reciprocals) When the reference angle is included in the description the ratios can be defined in terms of the reference angle C.

Definition 49 *The **sine of angle C**, written* $\sin C$, *is the ratio between the side opposite to* $\angle C$ *and the hypotenuse of* $\triangle ABC$,

$$\sin C = \frac{opposite \, side \, to \, \angle C}{hypotenuse} = \frac{\overline{AB}}{\overline{BC}}. \tag{3.9}$$

Definition 50 *The **cosine of angle C**, written* $\cos C$, *is the ratio between the side adjacent to* $\angle C$ *and the hypotenuse of* $\triangle ABC$,

$$\cos C = \frac{adjacent \, side \, to \, \angle C}{hypotenuse} = \frac{\overline{AC}}{\overline{BC}}. \tag{3.10}$$

Definition 51 *The **tangent of angle C**, written* $\tan C$, *is the ratio between the side opposite to* $\angle C$ *and the side adjacent to* $\angle C$,

$$\tan C = \frac{opposite\ side\ to\ \angle C}{adjacent\ side\ to\ \angle C} = \frac{\overline{AB}}{\overline{AC}}. \tag{3.11}$$

Remark 52 *Since* $\triangle ABC$ *is similar to* $\triangle DEF$ *then* $\angle C = \angle F$ *and so*

1. $\frac{\overline{AB}}{\overline{BC}} = \frac{\overline{DE}}{\overline{EF}}$ *which means* $\sin C = \sin F$.

2. $\frac{\overline{AC}}{\overline{BC}} = \frac{\overline{DF}}{\overline{EF}}$ *which means* $\cos C = \cos F$.

3. $\frac{\overline{AB}}{\overline{AC}} = \frac{\overline{DE}}{\overline{DF}}$ *which means* $\tan C = \tan F$.

This implies that as long as the triangles are the same "shape," that is, the angles are the same then the ratios of sine, cosine and tangent of the reference angle will be the same. Engineers and architects realized that if they had a table of values for the sine, cosine and tangent of an angle then they could use these values in whatever proportion was needed. In this way the concept of ratio and proportion can be used to make a special comparison and help solve the problem at hand.

We do not need to refer to such tables if we have a modern calculators. Most modern calculators have sine, cosine and tangent buttons as standard functions. It is simply a matter of "plugging" into the calculator the correct angle measure.

Exercise 53 *A tree of unknown height stands 10 meters from a nearby house. Civil engineers have drawn up plans to widen a passing roadway which requires the removal of the tree. Although the standard procedure is to remove sections of the tree starting from the top and working their way down, the engineers want to be sure nothing goes wrong. This means they need to measure the height of the tree to be sure that the tree does not hit the house if it happens to all fall at once. Standing at the base of the house the engineers determine that the angle from the base of the house to the top of the tree is about* $42°$.

1. *As a class exercise determine if there is a danger of the tree hitting the house if it falls all at once?*

2. *Of course, in order to use trigonometry in this way a method for measuring the angles is needed. This can be accomplished with a simple protractor, a piece of string and a small weight (perhaps a fishing weight). Attach one end of the string to the center of the protractor and attach the weight to the other end of the string (see diagram 3.12). Discuss how this modified protractor can measure the angle from the base of a building to the top of the building from a fixed distance from the building.*

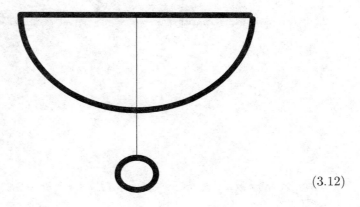

(3.12)

Protractor with a Weight Attached

3. *Use trigonometry to once again measure the height of the glass entrance to MCT. How does this trigonometric measurement compare to the previous answer you obtained? Explain.*

Remark 54 *If we chose $\angle B$ instead of $\angle C$ as the reference angle then*

1. $\sin B = \cos C$. *Explain why.*

2. $\cos B = \sin C$. *Explain why.*

3. *the tangent of $\angle B$ is the reciprocal of the tangent of $\angle C$,* $\tan B = \frac{1}{\tan C}$. *Explain why.*

Therefore, it is not necessary to record the sine, cosine and tangent of both acute angles in a right triangle. The trigonometric ratios of one acute angle in a right triangle are the reciprocals of the ratios of the other acute angle in that triangle.

3.3.4 Problems

1. Search the internet for a map of your area and zoom to within 100 square miles. Print out this map to use as a guide. Use the concepts of relative density and/or relative frequency to answer the following.

   (a) Estimate the degree to which "urban sprawl" has occurred within your area.

   (b) Determine the ratio between "urban" and "rural" regions in the map.

   (c) Determine the ratio of "urban" area to the total area of the map.

   (d) What does this tell you about your area?

   (e) What factors in our experiment could affect the results.

2. Repeat problem 1 with a map of an area that seems to be more urban. Repeat the problem again with an area that seems to be less urban. Select areas about which you have little first-hand knowledge.

3. Geologists and ecologists can use a measure of the amount of water surface area to determine whether or not the region is experiencing drought conditions. If a "standard" water surface area is established during normal rainfall conditions, this standard can be compared to measured conditions during times of abnormal rainfall. Scientists rely on aerial photographs to determine the amount of water surface area. For the purposes of this exercise, search the internet for a map of your area and zoom to 100 square miles. Print this map and use it as your aerial photograph. Use the concepts of relative density and/or relative frequency to answer the following.

   (a) Estimate the current amount of water surface area in your region.

   (b) Determine the ratio between water surface area and total area.

   (c) Even though we do not know the standard water surface area for the region, does this information tell us something about water conditions for the area?

   (d) Suppose a current aerial photo showed substantially less water surface area, say 20% less. Can we conclude that there is definitely a water shortage? What factors other than water surface area could affect our results?

4. Repeat problem 3 with a map of an area that seems to have more water surface area. Repeat the problem again with an area that seems to have less water surface area. Select areas about which you have little first-hand knowledge.

5. Coffee is brewed by allowing hot water to percolate through ground coffee beans. At many supermarkets, you will find coffee grinding machines that grind coffee to various "grinds."

(a) What is the difference between the different "grinds"?

(b) How would different "grinds" affect the final coffee product assuming the same type of beans are used for each different "grind"?

6. Tea is generally steeped rather than percolated; that is, the tea leaves are contained in a bag or strainer of some sort and soaked in hot water until the desired strength is obtained.

   (a) Why might it be undesirable then to percolate tea as we do coffee?

   (b) What could be done to the tea leaves to allow you to percolate them rather than steep them?

   (c) What does this tell you about the less than enthusiastic acceptance of "coffee bags" recently introduced in the marketplace?

7. When a lot is purchased for the purpose of building a house, the land is usually subjected to a perc test. The test procedure starts by digging a hole and filling it with water. The time it takes for the water to drain from the hole is measured. Suppose the test was conducted several times with the following results. What can be said about the porosity of the soil at the various test sites? Assume all the water drained through the bottom of the hole and each hole was filled to the top before left to empty.

Site number	Length (feet)	Width (feet)	Depth (feet)	Time to drain (hours)
1	6	4	3	46
2	3	4	2	13
3	4	4	3	22
4	2	3	2	9
5	3	3	3	11
6	2	2	1	1/4

8. For each site in the previous problem, conjecture what the soil texture might be like. What would be a purpose for conducting such a test? Is it possible to have a particular application where the porosity of the soil is too high? Explain.

9. When drilling a well, one concern is to drill the well deep enough so that the well will fill more quickly than the water will be used. This is measured as the flow capacity of the well. Most wells used to supply water for a single house have a rated capacity of 200 gallons per hour meaning you can pump water out at two hundred gallons per hour and not empty the well.

   (a) Two hundred gallons per hour seems like a lot of water. How does this translate into water usage inside the average home?

   (b) Assuming you will not drill into an underground pocket of water (e.g. an underground lake or stream), why would drilling the well deeper increase the flow capacity of the well?

(c) What soil factors would affect how deep the well would need to be drilled?

(d) If we drill a well 100 feet deep and have a flow capacity of 100 gph, is it safe to assume that drilling another 100 feet will yield a flow capacity of 200 gph? Explain.

10. By drilling the well deeper, we also increase the reserve capacity (the total amount of water in the well). Assuming an average family of four uses about 500 gallons of water per day, how deep would we need to drill a nine inch diameter having a flow capacity of 200 gallons per hour to ensure the well would not run dry? (Hint: some assumptions must be made here. The reserve capacity is the volume of a nine inch diameter column of water of a certain height. Assuming this column re-fills at 200 gph, how big must the column be to ensure not running out of water? We must also consider how the water is used. Generally, water usage in a home is not uniform but rather concentrated in the morning and evening hours.)

11. What is the purpose of lining a landfill site with clay?

12. The United States is the last industrialized country to use the English measurement system, the other countries of the world having adopted the metric system. Conversion to the metric system in the U.S. has been slow; however, it costs U.S. companies hundreds of billions of dollars a year to conduct international trade because of the need to maintain measurements and packaging for both systems. Some of the more common conversions are 1 kilometer $\approx$ 0.6 miles, 2.54 centimeters $\approx$ 1 inch, 1 kilogram $\approx$ 2.2 pounds and 1 litre $\approx$ 0.94 quarts. Following are some common measurements you see regularly. Identify the source of the measurement and perform the conversion.

(a) Convert...

   i. 100 miles to kilometers.
   ii. 26 miles to kilometers.
   iii. 1.5 miles to meters.
   iv. 2.65 miles to meters.
   v. 100 yards to meters.
   vi. 15 feet to meters.
   vii. 9 inches to centimeters.
   viii. 36 inches to meters.

(b) Convert...

   i. 15 kilometers to miles.
   ii. 10 kilometers to miles.
   iii. 5 kilometers to miles.
   iv. 800 meters to miles.

      v. 400 meters to yards.

     vi. 10 centimeters to inches.

    vii. 100 centimeters to feet.

   viii. 1 meter to inches.

    ix. 10 meters to feet.

     x. 100 meters to yards.

(c) Convert...

      i. 120 pounds to kilograms.

     ii. 180 pounds to kilograms.

    iii. 5 pounds to kilograms.

    iv. 1 pound to grams.

     v. 1 ton to kilograms.

    vi. 44,000 pounds to kilograms.

(d) Convert...

      i. 30 grams to ounces.

     ii. 100 grams to pounds.

    iii. 100 kilograms to pounds.

    iv. 1,000 kilograms to tons.

     v. 1,000,000 kilograms to tons.

(e) Convert...

      i. 50 gallons to litres.

     ii. 1 gallon to litres.

    iii. 0.5 gallon to liters.

    iv. 1 ounce to millilitres.

     v. 16 ounces to litres.

    vi. 20 ounces to litres.

(f) Convert...

      i. 4 litres to gallons.

     ii. 3 litres to quarts.

    iii. 2 litres to quarts.

    iv. 1 litre to ounces.

     v. 0.5 litres to ounces.

    vi. 50 millilitres to ounces.

13. Research project. A common problem that occurs in the English measurement system that does not occur in the metric system arises from the fact that in the English system, measurement of a particular quantity may be conducted in different unrelated ways. For example volume may be measured in cubic linear measurements but also in ounces, quarts and

gallons. In the metric system, liters are related to cubic centimeters in a very specific way whereas in the English system, gallons bear little relation to cubic feet. Nonetheless, conversions between these various unrelated systems is often necessary. Research conversion factors for the following conversions.

(a) Cubic feet to gallons.

(b) Cubic inches to gallons.

(c) Miles to nautical miles.

(d) Acres to square miles.

14. Research Project. Depending on the species of tree, it is often possible to predict the height of a tree by measuring the diameter of its base. Identify a particular species of tree and test this claim by measuring the diameter of the base of the trunk and the height of several specimens to calculate the ratio. Perform this experiment for several species. Can you find any pattern in the ratios among various species? For example compare data collected on various different evergreens to each other. Do the same for several deciduous species.

15. Research Project. Find out what the Golden Ratio is and what its historical importance is in art, architecture and design. Write a short paper and/or prepare a presentation to give to your classmates.

3.4   Inverse Proportions

Thus far this chapter has discussed a comparison of quantities known as ratio and proportion. That is, the ratios $\frac{a}{b}$ and $\frac{c}{d}$ are in proportion if

$$\frac{a}{b} = \frac{c}{d}. \tag{3.13}$$

Another way to describe the relationship between the quantities in 3.13 is to say that if the values of c and d remain constant, say $\frac{c}{d} = k$, k a constant, then *the values a and b are in direct proportion.*

$$\frac{a}{b} = k.$$

That is, if the value of a were to increase, then the value of b would also have to increase a proportional amount so that the fractional value $\frac{a}{b}$ will remain constant.

This treatment implies that the values of c and d are known and we wish to find the value of either a or b given that we also know the other value. However, there are applications that require the use of the relationship between a and d rather than between a and b. In this case proportion 3.13 is written

$$ad = k,$$

where k is a nonzero constant value, or

$$a = \frac{k}{d}.$$

Note that if the value of d were to increase, then the value of a must decrease. Similarly, if the value of d were to decrease, then the value of a must increase. In this case the values a and d are said to be *inversely proportional.* Note also that $k \neq 0$ implies that neither a nor d can be zero.

Definition 55 *Two values a and d are **inversely proportional** if*

$$ad = k$$

for some nonzero constant k.

3.4.1   Highway Safety

There are many applications of inverse relations in everyday life. For example, the following exercise demonstrates that over a fixed distance, speed and time are inversely proportional.

Exercise 56 *Many people believe that by driving faster they will save alot of time. Although greater speed will save some time this exercise demonstrates that at greater speeds the amount of time saved for a fixed increase in speed diminishes very rapidly. Suppose a driver plans to travel a distance of 20 miles along a stretch of interstate highway. Use complete sentences in your explanations and show all necessary calculations.*

1. *Recall the formula that relates distance D, time T and rate R (that is, speed) with each other and write it in the box below.*

2. *Use this formula to fill in the following chart with data. The chart records how long it will take to travel 20 miles at the indicated speeds. Note that the distance remains constant at 20 miles.*

RATE (SPEED)	40	45	50	55	60	65	70	75	80	85
TIME										

You can rely on a calculator to perform all necessary calculations.

3. *One method of analyzing the results in the previous table is graphically. To do this we must first identify which values represent the domain and which values represent the range. Explain which (RATE or TIME) is the domain and which is the range.*

4. *Plot the data on the grid below. Label the domain and range axes appropriately.*

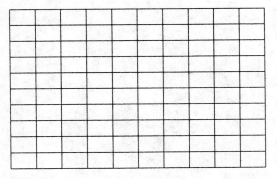

(a) *Examine the pattern of points closely. Describe what type of curve appears.*

(b) *Make a prediction about how long it would take to travel 20 miles if you were traveling at 90mph. 35mph. 100mph. Explain.*

(c) *Explain what happens to the time saved when the driver increases from 40mph to 45mph as compared with the increase from 75mph to 80mph). Is it always worth the time saved to travel faster?*

The relationship between the speed of a car and the time saved is of great concern to highway safety analysts. Some of the exercises at the end of this section outline the types of analyses performed for evaluating optimal speeds for cars to travel along highways given travel conditions. Other inverse proportion relationships are illustrated in the exercises as well.

3.4.2 Problems

1. A local mall is constructing shops inside the mall to be leased out. In general, all the shops will be rectangular and be constructed to specific sizes with regard to floor area. For this exercise, we will consider constructing rectangular shops having 6,400 square feet where x and y represent the dimensions of the store.

   (a) List at least five different possible sets of dimensions the store could have. Verify that the floor space is 6,400 square feet in each case. Plot the data points you've found on a graph. Sketch a curve that fits the data.

   (b) Based on your analysis in part 1a, write down an equation that expresses the relationship between the dimensions of the store and the square footage.

   (c) How does the value of y change as you increase the value of x? decrease the value of x?

   (d) Solve the equation from part 1b to find a formula that yields the corresponding value for y assuming you are given a value for x. What does this formula say of the relationship between x and y?

   (e) For the formula in part 1d, what is the possible range of values for x and y; that is, what are the domain and range of your function? Does the graph of this function correspond to the curve you sketched in part1a?

   (f) How do these values for domain and range relate to the original problem? Can you describe a more practical set of values for the domain and range?

2. Wire of a given diameter is manufactured by pulling thicker wire though dies that compress the diameter while lengthening the wire. The total volume of wire material remains constant. To start the process, and ingot of copper is cast that contains the appropriate weight (and hence volume of materials) as the finished spool of wire. This ingot is pressed through rollers of ever decreasing diameter until its diameter is small enough to pull through the wire dies.

   (a) Assuming we wish to obtain a spool of wire 300 meters long and 1 millimeter in diameter, what volume of copper is required to make the spool.

   (b) What must the length of the ingot be to construct such a wire if the initial diameter is 100 millimeters?

   (c) What is the length of such an ingot after being compressed to a diameter of 85 millimeters?

(d) The wire must be reduce in diameter to at most 5 millimeters before it can be pulled through the available dies. At that point each die will reduce the diameter 0.5 mm until it reaches the desired diameter. Prior to that point, the rollers available can reduce the diameter by 5 mm. Complete a table showing the diameter and length of the wire after each reduction.

(e) Write down a formula that computes the length of the wire l for a given diameter d. What does this formula tell you about the relationship between l and d? Graph your formula.

(f) Specify the domain and range of your formula. How long will the wire be if the diameter is reduced to 0.5mm? to 0.1mm?

3.5   Summary

This chapter has outline many of the mathematical models used to compare two quantities. Ratio and proportion can be used for both direct and inverse relationships. Although quite simple mathematically, these models have many practical applications. When studying the characteristics of a situation, it is often impossible or impractical to observe the data directly. By comparison to known situations (e.g. samples), insight can be gained as to the characteristic of the data that is otherwise difficult to obtain. The examples outlined in this chapter illustrate the importance of being able to utilize these models effectively. Applications such as estimating population sizes, determining the distribution of trees in a forest, porosity of soil samples and highway safety are important to understanding the world around us.

Chapter 4

Quadratic Function Models

4.1 Introduction

A good rule of thumb to use when developing models is "keep it simple." Always use the simplest, most accessible model available that satisfies the conditions of the problem. The simplest model is the linear model. In the previous chapter we studied many practical examples of applications of linear models. There are many examples, however, where linear models are not accurate enough. For example, recall from section 1.3.3 the example of wind velocity and the force it generates.

Term(n)	1	2	3	4	5	6	7	8
Wind Velocity$(W(n))$	30	35	40	45	50	55	60	65
Force(lb/sq ft)	4.5	6.125	8.0	10.125	12.5	15.125	18.0	21.125

Although the wind velocity is easily represented by a linear expression, $W(n) = 5n + 25$, the force is clearly not progressing in a linear fashion. In fact, using the sequencing technique from chapter 1 we can fill in the following table with the appropriate differences to recognize a pattern.

Term (n)	Force $(f(n))$	Difference$(f(n) - f(n-1))$	Second Difference
$n = 1$	4.5	Blank	Blank
$n = 2$	6.125	1.625	Blank
$n = 3$	8.0	1.875	0.25
$n = 4$	10.125	2.125	0.25
$n = 5$	12.5	2.375	0.25
$n = 6$	15.125	2.625	0.25
$n = 7$	18.0	2.875	0.25

$$(4.1)$$

115

Notice that the second difference between the terms is constant. Thus, the first difference is growing in a linear fashion. That is,

$$\text{First Difference} = 0.25n + 1.625 \tag{4.2}$$

This means that the original force expression can be obtained from 4.2 and is of the form

$$F(n) = n(0.25n + 1.625) + (\text{a constant}). \tag{4.3}$$

This would imply that $F(n)$ is quadratic in nature. There are many ways of determining what expression 4.3 should be. In the next section we will study one of these methods. In succeeding sections we will discuss additional applications of quadratic models.

4.2   Finding a Quadratic Model to Fit Data

4.2.1   Finite Differences Applied to Quadratic Data

We know from both the table 4.1 and from the expression 4.3 that the model representing the force asserted by wind of a given velocity is quadratic. The concept of **finite differences** is a direct way to find this quadratic model. Since we know the model is quadratic it must be of the form

$$F(n) = an^2 + bn + c.$$

So if we determine the values of $a, b,$ and c we will then have the quadratic model. To do this we construct a second *symbolic* table analogous to the *numerical* table 4.1 above.

Term	Force	Numerical	First	Num.	Second	Num.
(n)	$f(n) = an^2 + bn + c$	Value	Difference	Value	Diff.	Value
$n = 1$	$a(1)^2 + b(1) + c$	4.5	Blank	Blank	Blank	Blank
$n = 2$	$a(2)^2 + b(2) + c$	6.125	$3a + b$	1.625	Blank	Blank
$n = 3$	$a(3)^2 + b(3) + c$	8.0	$5a + b$	1.875	$2a$	0.25
$n = 4$	$a(4)^2 + b(4) + c$	10.125	$7a + b$	2.125	$2a$	0.25
$n = 5$	$a(5)^2 + b(5) + c$	12.5	$9a + b$	2.375	$2a$	0.25
$n = 6$	$a(6)^2 + b(6) + c$	15.125	$11a + b$	2.625	$2a$	0.25
$n = 7$	$a(7)^2 + b(7) + c$	18.0	$13a + b$	2.875	$2a$	0.25

$$\tag{4.4}$$

In fact, we do not need the entire table. Since there are only three unknowns $(a, b,$ and $c)$ we need to form only three equations. So we will choose the three easiest equations. The equations are formed by equating each of the numerical values in table 4.4 with the corresponding symbolic values in table 4.4.

$$
\begin{aligned}
a + b + c &= 4.5 \\
3a + b &= 1.625 \\
2a &= 0.25
\end{aligned}
$$

Solving this system of three equations and three unknowns (by substitution), we have

$$a = 0.125,$$
$$b = 1.25 \text{ (substituting } a = 0.125 \text{ into second equation)},$$
$$c = 3.125 \text{ (substituting } a = 0.125 \text{ and } b = 1.25 \text{ into third equation)}.$$

Therefore, the force has quadratic representation

$$f(n) = 0.125n^2 + 1.25n + 3.125.$$

Exercise 57 *Use the quadratic model to determine the force generated by a wind gust of 120 miles per hour. (Note: be careful what value is substituted in for n).*

Exercise 58 *As stated earlier, no model is perfect. There is always some error in the application of mathematical modeling. Identify some ways in which this model may be in error.*

4.2.2 Finite Differences Applied to Quadratic Graphs

This technique of finite differences can be adapted and used to find the analytic expression of a parabolic curve. The curve associated with a quadratic function is called a parabola.

Definition 59 *Given a line l and a point P(h, k) not on l the parabola associated with point P(h, k) and line l is the locus of points (x, y) such that the distance from (x, y) to P is equal to the distance from (x, y) to l.*

Here is an exercise to demonstrate this:

Exercise 60 *You were asked to test drive the new Rusty Motors electric car to see how it handled on interstate highways. The first graph below represents your variable velocity (range) over the 8-hour (time - domain) test session; the second graph below represents your actual position/distance (range) on the highway over that time (domain).*

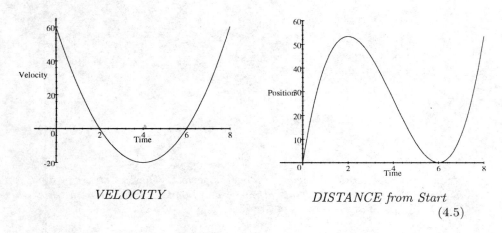

VELOCITY   DISTANCE from Start

(4.5)

1. What was the furthest you traveled from the starting point? At what times did you achieve that distance?

2. Explain what may have happened between the hours of 2 and 6.

3. *Notice that the first graph, the velocity graph, is parabolic in shape. This tells us that in searching for an analytic expression representing the velocity of the car we should be looking for one that is quadratic. That is, $v(t) = at^2 + bt + c$ will represent the velocity of the car. To find a,b and c, we need only read at three points off the graph, however, we will find the first and second differences of the data in order to demonstrate the technique. So, using our best judge as to the points on the curve (pay special attention to the signs that are used),*

Point	Curve	Value	First	Diff.	2nd	Diff.
(t)	Numerical	Symbolic	Num.	Sym.	Num.	Sym.
$t=0$	60	c	Blank	Blank	Blank	Blank
$t=2$	0	$4a+2b+c$	-60	$4a+2b$	Blank	Blank
$t=4$	-20	$16a+4b+c$	-20	$12a+2b$	40	$8a$
$t=6$	0	$36a+6b+c$	20	$20a+2b$	40	$8a$
$t=8$	60	$64a+8b+c$	60	$28a+2b$	40	$8a$

It helps in this process to pick points that are "obvious to read" (that is, we may need to "doctor" the data a little). Setting the Numerical equal to the Symbolic in each case we have the system of equations:

$$8a = 40; 4a + 2b = -60; c = 60 \qquad (4.6)$$

This quickly can be simplified to

$$a = 5; b = -40; c = 60$$

Hence, the quadratic model that represents the parabolic curve in the first graph in 4.5 is

$$v(t) = 5t^2 - 40t + 60. \qquad (4.7)$$

Once the analytic model is found it can be used to answer additional questions about the problem.

Exercise 61 *Use the velocity function model 4.7 to answer each of the following.*

1. *What is the velocity of the car at the seventh hour of the test drive? Does your answer agree with the graph 4.5?*

2. *When during the test is the car traveling at 25 mph? Explain.*

3. *What is the velocity of the car at the third hour of the test drive? Does the fact that the answer is negative make sense? Explain. Does your answer agree with the graph 4.5?*

In each of the previous examples the finite differences in the data worked out exactly. This is not usually the case when working with real data. The next section demonstrates how to use the technique of finite differences when the data is not exact.

4.2.3   Finite Differences Applied to Quadratic Regression

Previously we defined regression as the art of fitting a curve to a given data set. In the previous sections the quadratic models fit the data exactly. That is, if we were to plot the points and the curve simultaneously on the same graph the curve would pass right through each data point. In reality this occurs very infrequently when dealing with data sets collected in a statistical study. For example, the table below represents the population of the United States in the census years from 1950 to 1990. The parenthetical numbers in the second column (numerical population) are the true population census numbers obtained from the national census bureau. When plotted on a graph they "appear" to take the form of a parabola but do not all fall directly onto one specific parabola. This makes sense in that populations do not grow "smoothly" over time. There are all sorts of unpredictable events that can slightly alter the amount. However, the general pattern is that of a parabola. By studying the path of the data points it appears that the second set of population numbers in column two are close enough to the actual values and they fall along the parabola. Thus to find a quadratic model for the U.S. population growth we will use these numbers.

Year	Population Num.	(millions) Sym.	First Num.	Diff. Sym.	2nd Num.	Diff. Sym.
(x=1)1950	151.3≈156					
(x=2)1960	179.3≈176					
(x=3)1970	203.3≈200					
(x=4)1980	226.5≈228					
(x=5)1990	248.7≈260					

$$(4.8)$$

1. Fill in the appropriate values in each column in table 4.8 above.

2. Set corresponding values (numeric=symbolic) equal to each other in order to determine a quadratic that models the U.S. population.

3. Use the quadratic model to make a prediction about the population in the years 2000 and 2030.

4. Comment on the accuracy of this quadratic model. What kinds of possible errors could occur and where? Which prediction, the one for the year 2000 or the one for the year 2030, are you more confident is a "good" estimate?

4.2.4 Problems

1. As a representative of the State Environmental Protection Agency you are asked to monitor the amount of waste being delivered on a yearly basis to the Lucky County Landfill Site (LCLS). You are given the following data:

Term (n)	Millions of Tons $(f(n))$	Diff.$(f(n) - f(n-1))$	2nd Diff.
$1988(n = 1)$	523.25	Blank	Blank
$1989(n = 2)$	528		Blank
$1990(n = 3)$	534.25		
$1991(n = 4)$	542		
$1992(n = 5)$	551.25		
$1993(n = 6)$	562		
$1994(n = 7)$	574.25		

$$(4.9)$$

(a) Fill out the numerical table and equate the corresponding values to the symbolic table. Find a function model for the number of tons of waste deposited at the LCLS over the years 1988 to 1994. Explain your calculations.

(b) Use the model from part a to make a prediction as to how much waste will be handled by LCLS this year and five years from now. Which of these two predictions would you expect to be more accurate? Explain.

(c) Identify some ways in which this model could be in error thus throwing off your prediction.

2. The rise and fall of stock values is always in the news. The graphs below represent the performance of a stock during a typical trading day on a Stock Exchange. Graph 1 below represents the rate at which that stock rose and fell in cents per hour (range) vs. the hourly time (domain) and graph 2 below represents the actual price of that stock in dollars (range) at those times (domain). The time scale represents the time from the opening bell $(t = 0)$ to the closing bell $(t = 8)$.

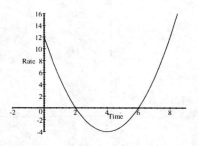

The RATE of price change

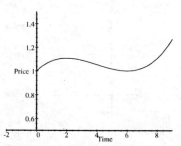

The PRICE at time t

(a) At what times during the day was the stock rising? falling? Explain using the RATE graph above.

(b) At what times during the day was the rate zero? Explain using the RATE graph above. What was the price of the stock at those times? Explain using the PRICE graph above.

(c) What was the price of the stock when the day began? when the day ended? Explain using the PRICE graph above.

(d) What was the price of the stock when it started to rise again? Explain using both graphs above.

(e) As an investor, when would have been the best times to buy the stock? to sell the stock? Explain.

(f) Although the graphs above are very descriptive it is often beneficial to use an analytic expression to describe the rate at which the stock changes. Study the graph for a moment and see if you can discover an analytic expression that fits the data. (Hints: What is the shape of the curve? Where does the curve cross the x-axis?)

(g) Use the analytic model found in part f. to answer each of the following. Check your answers by comparing them to the graph.

    i. At what rate is the stock rising at the fifth hour of the day? Explain.

    ii. When is the stock rising at 5 cents per hour? Explain.

    iii. When is the stock falling at 4 cents per hour? Explain.

3. Scientists are concerned about the effects of modern civilization on a local species of bird. Data was collected that appears in the graphs below. The two graphs below represent a year of measuring the bird population. Graph 1 below represents the rate at which the population changed in hundreds of birds per month (range) vs. the month (domain) and graph 2 below represents the actual population in hundreds of birds (range) over those months (domain). The time scale, measured in months, represents the time from the first month of the month ($t = 0$) to the last ($t = 12$).

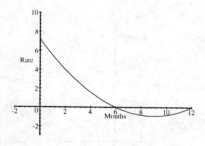

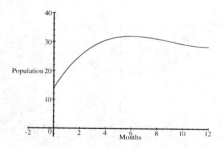

The RATE Pop. changed   The POP. on month t

(a) At what times during the year was the population rising? falling? Explain using RATE graph above.

(b) At what times during the year was the population change zero? Explain using RATE graph above. What was the population at those times? Explain using the POP. graph above.

(c) What was the population when the year began? Explain using the POP. graph.

(d) What was the population when the year ended? Explain using the POP. graph.

(e) Again the graphs above are very descriptive, but it is beneficial to use an analytic expression to describe the rate at which the population changes. Study the graph for a moment and see if you can discover an analytic expression that fits the data. (Hints: What is the shape of the curve? Where does the curve cross the x-axis?)

4. Scientists are concerned that warmer weather trends will create conditions that increase the number of insects in affected regions. The graphs below represent measuring an insect population over three months. Graph 1 below represents the rate at which the population changed (range) vs. the week (domain) and graph 2 below represents the actual population (range-in thousands of insects) during those weeks (domain). The time scale represents the time from the first week of the study ($t = 0$) to the last ($t = 12$).

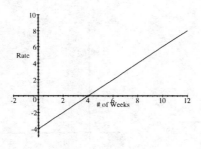

The RATE Pop. changed   The POP. on week t

(a) At what times during the study was the population rising? falling? Explain using the RATE graph.

(b) At what times during the study was the population the least? greatest? What was the population at those times? Explain using the POP. graph.

(c) What was the population when the study began? ended? Explain using the POP. graph.

(d) Find an analytic expression to describe the rate at which the population changes (the first graph). (Hints: What is the shape of the curve? Where does the curve cross the x-axis?)

(e) Use the analytic model in part d. to determine the rate the bird population is changing at the sixth month and the eleventh month.

(f) Find an analytic expression to describe the size of the population (the second graph). (Hints: What is the shape of the curve? Where does the curve cross the x-axis?)

(g) Use the analytic model in part e. to determine the size of the bird population at the sixth month and the eleventh month.

5. Recall the linear example we did on traveling the highway and speed. You were asked to monitor the trip of one of the truck drivers of a distribution company. In that example we assumed that the driver maintained a constant speed over long distances. Of course, this is not very realistic. We know from our own experiences that drivers rarely maintain the same velocity over long distances. We are constantly speeding up or slowing down. The two graphs below represent a more realistic case involving the truck driver. Graph 1 below represents the variable velocity (range) at which the driver was traveling over time (domain) and graph 2 below represents the actual distance (range) she traveled over that time (domain). The time scale, measured in hours, represents the time from the beginning of the trip ($t = 0$) to the end of the trip ($t = 6$).

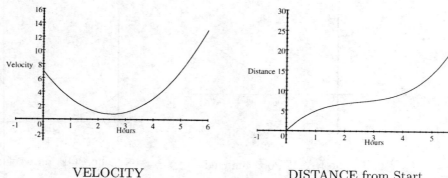

VELOCITY   DISTANCE from Start

(a) At what times during the trip was she speeding up? slowing down? Explain using the VELOCITY graph.

(b) When did she begin to speed up? Explain using both graphs 1 & 2.

(c) What was her speed at the beginning of the trip? at the end? Explain using the DISTANCE graph.

(d) How long was the trip all together?

(e) Again study the velocity graph for a moment and see if you can discover an analytic expression that fits the data. (Hints: Note that this graph does not cross the x-axis. What does this mean as to how you will find the analytic model?)

(f) Use the analytic model in part e. to determine the driver's speed at the first and fifth hours of the trip.

(g) Repeat questions a through f using the following graphs for the driver's velocity and distance of the trip.

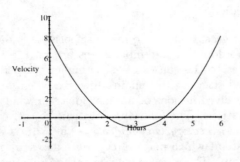 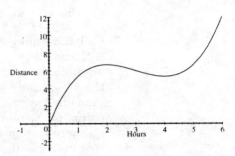

VELOCITY   DISTANCE from Start

6. Vacation travel has always been an essential part of the American Dream. Each year millions of Americans plan vacations. One mode of transportation that has been growing in popularity in recent years is air travel. However, with the increased popularity of air travel comes an increase in the headaches and problems of air travel as well. One indication of this is the increase since 1990 of the number of passengers bumped from U.S. based flights. The chart below represents the number of passengers bumped from U.S. flights (in millions) since 1990 (Source: Newsweek)

Year	Pass. Bumped (millions) Numerical	Symbolic	First Num.	Diff. Sym.	Second Num.	Diff. Sym.
$(x=0)$ 1991	(0.65) _ _ _ _					
$(x=2)$ 1993	(0.7) _ _ _					
$(x=4)$ 1995	(0.85) _ _ _					
$(x=6)$ 1997	(1.05) _ _ _					

(4.10)

(a) Estimate some values in the first column of 4.10 for the number of passengers bumped so that a reasonable constant second difference results. Fill out the numerical table , equate the corresponding values to the symbolic table. Find a quadratic function model for the number of passengers bumped over the years 1991 to 1997. (Use the finite differences technique.) Explain your calculations.

(b) Use the model generated in part a above to make a prediction as to how many passengers were bumped from flights in 1999. In the year 2000. Explain.

(c) Use the model generated in part a above to make a prediction as to how many passengers will be bumped from flights five years from now. Would you expect this answer to be more or less accurate at predicting the number of "bumped" passengers than the answers found in part b.?Explain.

(d) Identify some ways in which this model could be in error thus throwing off your prediction.

7. Global warming is of great concern to many scientists around the world. Although there is no doubt that temperatures are on the rise, there is much disagreement as to the causes of those higher temperatures. It is the job of every good scientist to remain objective when studying data, shunning any preconceived notions or popular ideas for what is the truest nature of the observations. The following data can be extrapolated from the website *http://www.nsc.org/ehc/guidebks*. This site is one of many that discusses the rate at which major greenhouse gases are increasing in the earth's atmosphere. These gases include carbon dioxide (CO_2, from fossil fuel combustion & deforestation), methane (CH_4, from fossil fuel production, rice fields, cattle, landfills), nitrous oxide (N_2O, from nitrogenous fertilizers, deforestation, biomass burning, refrigerants, foams), and CFC-12 (from aerosol sprays, refrigerants, foams). The chart below shows the increase in the amount of methane in parts per million (ppm) in the atmosphere since the early part of this century.

| | Methane | (ppm) | First | Difference | Second | Diff. |
Year	Numerical	Symbolic	Numerical	Symbolic	Numerical	Symbolic
($x = 0$)1930	(1.10)____					
($x = 2$)1950	(1.21)____					
($x = 4$)1970	(1.40)____					
($x = 6$)1990	(1.72)____					

$$(4.11)$$

(a) Estimate some values in the first column of 4.11 for the amount of methane in the atmosphere in ppm so that a reasonable constant second difference results.

(b) Fill out the numerical table and equate the corresponding values to the symbolic table. Find a quadratic function model for the amount of methane in the atmosphere in ppm over the years 1930 to 1990. (Use the finite differences technique.) Explain your calculations.

(c) Use the model generated in part a above to make a prediction as to how much methane will be in the atmosphere by the year 2030. Explain.

    (d) Use the model generated in part a above to determine how much methane was in the atmosphere in the year 1910. Explain.

    (e) Identify some ways in which this model could be in error thus throwing off your prediction.

4.3   Multiplying Binomials (FOIL)

In the previous section we introduced a new, non-linear, model called the quadratic model ($f(x) = ax^2 + bx + c$). This model is more complex than the linear model studied in the previous chapter. As a result it possesses properties from the linear model. In this section we study some these properties. We gain a better understanding of what a functional model is telling us and the problem it represents by studying the properties of this functional model.

Both the linear and quadratic models are a part of a larger class of models called **polynomials**.

Definition 62 *A **polynomial** is an expression of the form $a_0 + a_1 x + \cdots + a_{n-1} x^{n-1} + a_n x^n$ where each a_i is an integer constant. Each part $a_i x^i$ is called a **term** of the polynomial. The **degree** of the polynomial is the value of n, the highest power in the polynomial. Both the terms and the degree of a polynomial are important in determining the "behavior" and properties of a polynomial model. A polynomial with one term is called a **monomial**; a polynomial with two terms is called a **binomial**; a polynomial with three terms is called a **trinomial**. A polynomial of degree one is called **linear**. A second-degree polynomial is called a **quadratic**.*

Example 63 *Here are some polynomial examples:*

1. $f_1(x) = x^5 - 3x^3 + 4x^2 + x - 8$; *the degree is 5 and the terms are* $x^5, -3x^3, 4x^2, x$ *and* -8.

2. $f_2(x) = x - 8$; *the degree is 1 (linear) and the terms are x and -8.*

3. $f_3(x) = 6x^2 - 3x + 5$; *the degree is 2 and the terms are $6x^2, -3x$ and 5.*

4. $f_4(x) = 4$; *the degree is 0 and the term is 4.. This polynomial is called a constant.*

We can perform simple operations on polynomial models such as adding or subtracting two polynomial models.

Example 64 *Add or subtract the following examples.*

1. $(3x^4 + 4x^3 - 5x - 6) + (2x^5 - 3x^3 + 2x^2 + 7x + 9) = 2x^5 + 3x^4 + x^3 + 2x^2 + 2x + 3$.

2. $(2x^5 - 3x^3 + 2x^2 + 7x + 9) - (3x^4 + 4x^3 - 5x - 6) = 2x^5 - 3x^4 - 7x^3 + 2x^2 + 12x + 15$.

Remark 65 *Note that we only combine together those terms with the same exponents. This is because $3x^4 + 2x^4$ is translated as 3 objects of the type "x^4" are added to 2 objects of the same type. Terms with the same variables and the same exponents are called **like terms**.*

General polynomial models will be studied in more depth in Chapter 5. For now we will concentrate primarily on linear and quadratic models. In particular, we will review the process of multiplying two binomials together. You may recall this process being referred to as FOIL (First, Outside, Inside, Last).

Example 66 *Multiply* $(x - 2)(2x + 3)$.

$$(x - 2)(2x + 3) = 2x^2 + 3x - 4x - 6 = 2x^2 - x - 6.$$

Example 67 *Multiply* $(3x + 4)(2x - 5)$.

$$(3x + 4)(2x - 5) = 6x^2 - 15x + 8x - 20 = 6x^2 - 7x - 20.$$

Example 68 *Multiply* $(x - 2)(2x + 3)$.

$$(x - 2y)(2x + 3y) = 2x^2 + 3xy - 4xy - 6y^2 = 2x^2 - xy - 6y^2.$$

Example 69 *Multiply* $(x + 3)^2$.

$$(x + 3)^2 = (x + 3)(x + 3) = x^2 + 3x + 3x + 9 = x^2 + 6x + 9.$$

In the next section we discuss the FOIL method from a different perspective, one that will prove useful to modeling applications.

4.3.1 The Tree and Rectangular Methods of Multiplying Polynomials

We begin with a review of how you multiplied polynomials when you were in high school.

Exercise 70 *Briefly describe the method you use for multiplying polynomials, for example* $(x^3 - 2x^2 + 3)(4x - 5)$.

We will take a closer look at this. You should be aware that there are other methods available for multiplying polynomials. Basically, when multiplying two polynomials together we must multiply each term in the first polynomial to each term in the second polynomial. There are many ways of organizing the terms so that we are sure to do all of the required multiplications.

Example 71 $(x^3 - 2x^2 + 3)(4x - 5) = 4x^4 - 13x^3 + 10x^2 + 12x - 15$

1. **Tree Method: The diagram in 4.12 below is referred to as a tree graph.** *This tree has two sections: the first section represents the trinomial $x^3 - 2x^2 + 3$, one stem for each term of the trinomial and the second section represents the binomial $4x - 5$, one copy of the binomial for each stem from the first section. Note that negative signs are placed with the corresponding terms.*

$$
\begin{array}{llll}
 & & 4x & = 4x^4 \\
 & x^3 & & \\
 & & -5 & = -5x^3 \\
 & & & \\
 & & 4x & = -8x^3 \\
 & -2x^2 & & \\
 & & -5 & = 10x^2 \\
 & & & \\
 & & 4x & = 12x \\
 & 3 & & \\
 & & -5 & = -15 \\
\end{array}
\tag{4.12}
$$

Each path from left to right represents one of the necessary multiplications of the polynomials above. By following each path we can perform the required multiplication and record the product to the right of the corresponding path. Once all the multiplications are completed we simply add down the column combining like terms where appropriate to obtain:

$$4x^4 - 13x^3 + 10x^2 + 12x - 15.$$

This method also is very helpful when multiplying three polynomials together:

As a class exercise find the product of each using the tree method.

(a) $(x - 2)^2(x^2 + 3)$

(b) $(x^2 - 3x + 2)(x - 4)(2x + 1)$

2. **The Rectangular Array Method:** *Another way to multiply two polynomials together and keep track of all the multiplications involved is to arrange the two polynomials, one horizontally - one vertically, in order to create a rectangular array of terms. In each corresponding space the product of that row and column term is placed. Using the previous example as a demonstration:*

	x^3	$-2x^2$	3
$4x$	$4x^4$	$-8x^3$	$12x$
-5	$-5x^3$	$10x^2$	-15

Note the terms in the corresponding entries agree with the terms from the tree method above.

4.3.2 Applying the Tree Method

There are several advantages to using the tree method. One such advantage is it allows us to better organize information to get a clear image of the problem. For example, suppose you work for a company that designs methods of security for major airports. Among the equipment you design and sell are metal detectors. However, due to the limitations of technology and materials the best design that you can come up with is a metal detector that only detects 80% of the objects that pass through it. This is obviously not good enough for your customers. Airport security cannot allow 20% of all metal objects to pass through their security system unchecked.

Problem 72 *How do you improve on this number? Spend some time in class discussing ways that more metal objects can be identified as they pass through the airport.*

One possible solution to this dilemma is to place several metal detectors together in series. That is, have passengers pass through several detectors one right after the other. It would be reasonable to ask if this improves the probability of detecting metal objects and if so by how much.

To demonstrate this we will make use of the tree method. Suppose we place three metal detectors in series. For each metal detector let x represent the case that a metal object is detected (success) and let y represent the case that a metal object is not detected (failure). The tree 4.13 shows all such possibilities.

$$
\begin{array}{ll}
x & = x^3 \\
y & = x^2y \\
x & = x^2y \\
y & = xy^2 \\
x & = x^2y \\
y & = xy^2 \\
x & = xy^2 \\
y & = y^3
\end{array}
\qquad (4.13)
$$

Note that the term x^3 represents the fact that all three machines detected the metal object, the term x^2y represents the fact that two of the three machines detected the metal object, etc. This tree of possibilities is identical to the tree that represents the product

$$(x+y)^3 = x^3 + 3x^2y + 3xy^2 + y^3. \qquad (4.14)$$

Thus the polynomial in 4.14 models the experiment of placing three metal detectors in series. In general, such a polynomial is referred to as a **generating function**. In this experiment it is only necessary that at least one metal detector successfully identifies the metal object. That is synonymous with the terms $x^3 + 3x^2y + 3xy^2$ in 4.14. If we let $x = .8$ for success and $y = .2$ for failure then the probability that at least one of the three detectors will spot the object is

$$(.8)^3 + 3(.8)^2(.2) + 3(.8)(.2)^2 = .992 \text{ or } 99.2\%.$$

This is a substantial improvement over the original 80%.[1]

There are many more ways in which the tree approach can be used to model problems. We will return to this technique and its applications in the next chapter.

---

[1] Actually this is not much different than what is already in place at most airports. When you pass through a metal detector you are actually passing through a "coil" of detectors. That is, you are passing through the same detector several times as you walk through the coil.

4.3.3 Problems

1. Use whichever method you think best to multiply each of the following polynomials.

   (a) $(x + 4)(x - 3)$

   (b) $(x - 7)(x - 5)$

   (c) $(2x + 1)(x + 8)$

   (d) $(2x + 3y)(x - y)$

   (e) $(x + y)^3$

   (f) $(x^2 - x + 1)(x + 2)$

   (g) $(x^2 - x + 2)(x^2 + 2x - 1)$

   (h) $(x^2 - xy + y)(x + 2y)$

   (i) $(2x - y)^2$

   (j) $(x^2 + x + 1)(x + 1)(x + 1)$

2. A medical test that tests for the ecoli bacteria is 70% effective. That is, 70% of the time it successfully identifies the presence of ecoli, 30% of the time it does not. The test will be done three times to improve the chances of detection.

   (a) Write a polynomial that represents the three trials using x to mean success and y to mean failure.

   (b) What percentage of the time will at least one of the tests be successful? Explain showing all calculations.

   (c) What percentage of the time will at least two of the tests be successful? Explain showing all calculations.

3. Space vehicles such as the space shuttle usually have several computers on board to handle the many complex calculations that need to be done. Suppose a space vehicle has 5 computers on board. The system is designed so that if one computer fails then the others will take over its responsibilities so the astronauts can be returned to earth safely. Let's say that there is a 10% chance that any one of the computers will fail (the actual percentage is much lower than this but we will use this percentage for our calculations).

   (a) Write a polynomial that represents whether the five computers will fail or not using x to mean success and y to mean failure. What are the values of x and y?

   (b) What percentage of the time will at least one of the computers remain operational? Explain showing all calculations.

   (c) What percentage of the time will at least one of the computers fail to remain operational? Explain showing all calculations.

   (d) What percentage of the time will all of the computers fail to remain operational? Explain showing all calculations.

4.4   Factoring Quadratic Models

In this section we will use the concept of factoring to help us analyze certain quadratic models. Before we look at any applications we will review the basic techniques of factoring quadratics.

4.4.1   Factoring Techniques.

There are several factoring techniques and several ways to go about factoring quadratics. In this section we will review some of them. Some general comments about factoring quadratics before we actually introduce the techniques.

Remark 73 *If you are familiar with another technique and prefer using that technique to the one discussed here then feel free to use that technique.*

Remark 74 *The word "quadratic" comes from the Latin word* quadratus *(past participle of* quadrare*) meaning "to make square." The relationship between* quadratus *and the latin word* quadrangulus *(four-cornered) may come from the geometric interpretation of area ($A = lw$). In the case where the length and width are unknown, written in terms of algebraic expressions ($c_0 x + c_1$) and ($c_2 x + c_3$), the product is the standard form of a quadratic expression.*

Remark 75 *The previous remark implies that given a quadratic expression $ax^2 + bx + c$, then if it factors it will always factor into the product of two linear expressions $ax^2 + bx + c = (c_0 x + c_1)(c_2 x + c_3)$. We will use this to our advantage.*

Remark 76 *The terms in a quadratic $ax^2 + bx + c$ are referred to as the square term, ax^2, the linear term, bx, and the constant term, c.*

Factoring Quadratics of the Form $x^2 + bx + c$

For a quadratic term of the form $x^2 + bx + c$ the square term has a coefficient of 1. This implies that this type of quadratic will factor into an expression of the form

$$x^2 + bx + c = (x + r_1)(x + r_2). \qquad (4.15)$$

We need only determine the values of r_1 and r_2. Recalling the FOIL procedure, in reverse

Remark 77 *Referring to the Outside and Inside terms of FOIL, the sum $r_1 + r_2$ is equal to the coefficient of the linear term b.*

Remark 78 *Referring to the Last terms of FOIL, the product $r_1 \cdot r_2$ is equal to the constant value c.*

Remark 79 *If $c > 0$ and $b > 0$ then r_1 and r_2 are both positive.*

Remark 80 *If $c > 0$ and $b < 0$ then r_1 and r_2 are both negative.*

Remark 81 *If $c < 0$ and $b > 0$ then $\max\{r_1, r_2\} > 0$ and $\min\{r_1, r_2\} < 0$.*

Remark 82 *If $c < 0$ and $b < 0$ then $\max\{r_1, r_2\} < 0$ and $\min\{r_1, r_2\} > 0$.*

Here are some exercises to work out in class.

Exercise 83 *Factor each of the following:*

1. $x^2 + 15x + 50$

2. $x^2 - 8x + 12$

3. $x^2 - 13x + 48$

4. $x^2 - 13x - 48$

5. $x^2 - 16x + 64$

Factoring Quadratics of the Form $ax^2 + bx + c$, $a \neq 1$

For a quadratic term of the form $ax^2 + bx + c$ the square term has a coefficient other than 1. This implies that this type of quadratic will factor into an expression of the form

$$ax^2 + bx + c = (s_1 x + r_1)(s_2 x + r_2). \qquad (4.16)$$

The process of determining the factors of this expression is slightly different. In this section we will outline a procedure that may be slightly different to the ones you may have seen in the past. We will demonstrate this technique through an example.

Example 84 *Factor* $6x^2 + 19x + 15$:

1. *Multiply the squared coefficient and the constant together, call the product* p; $p = ac$.

$$p = 6 \cdot 15 = 90$$

2. *Determine factors of p whose sum is the coefficient of the linear term, b;* $f_I + f_O = b$, $f_I \cdot f_O = p$.

$$10 + 9 = 19; \ 10 \cdot 9 = 90.$$

3. *Determine factors of f_O (for the Outside terms) and f_I (for the Inside terms) that when crossed produce the squared coefficient a and the constant term c;* $s_1 \cdot r_2 = f_O$, $r_1 \cdot s_2 = f_I$.

$$2 \cdot 5 \ = \ 10; \ 3 \cdot 3 = 9$$
$$2 \cdot 3 \ = \ 6; \ 5 \cdot 3 = 15$$

4. *The terms s_1, r_2, r_1, s_2 are the terms of the binomials.*

$$6x^2 + 19x + 15 = (2x + 3)(3x + 5).$$

Try some on your own.

Exercise 85 *Factor* $4x^2 - 22x + 30$

Exercise 86 *Factor* $10x^2 + 3x - 8$

Factoring Differences of Squares and Sums and Differences of Cubes

You may recall the special cases of factoring differences of squares and differences and sums of cubes. The techniques are repeated here for review.

Example 87 *Difference of squares:* $a^2 - b^2 = (a-b)(a+b)$. *Factor each of the following.*

    1. $25x^2 - 16y^2$

    2. $49x^6 - 36y^4$

Example 88 *Difference of cubes:* $a^3 - b^3 = (a-b)(a^2 + ab + b^2)$. *Factor each of the following.*

    1. $8x^3 - 27y^3$

2. $64x^6 - y^9$

Example 89 *Sum of cubes: $a^3 + b^3 = (a+b)(a^2 - ab + b^2)$. Factor each of the following.*

1. $8x^3 + 27y^3$

2. $64x^6 + y^9$

Factoring by Grouping

There is another common type of factoring. Although it does not involve quadratics we usually introduce it at the same time and this text will be no different. The technique of **factoring by grouping** is generally used on expressions of the form

$$ax + ay + bx + by.$$

This technique involves identifying some terms within the expression that have a common factor and then repeating the procedure twice.

Example 90 *Factor $3x - ay + 3y - ax$*

1. *Note that the first and third terms have a common factor of 3 and that the second and fourth terms have a common factor of $-a$. Therefore, rewrite the terms as*

$$(3x + 3y) + (-ax - ay).$$

2. *Factor out the common factor in each group.*

$$3(x + y) - a(x + y).$$

3. *Note that the entire expression now has a "common factor" of $(x + y)$. Factor this out of the entire expression.*

$$3x - ay + 3y - ax = (x + y)(3 - a).$$

Factoring is now complete.

Exercise 91 *Factor each of the following in class.*

1. $4x^2 + 4x + 5x + 5$

2. $4x^2 + 4x + 5x + 5$

4.4.2 Applications of Factoring to Models

Recall from the variable stock-rate example in section 4.1.2 that the analytic expression for the velocity at which we tested General Motors' new electric car was

$$v(t) = t^2 - 8t + 12.$$

Note that between the hours of $x = 2$ and $x = 6$ the VELOCITY is negative which means the car was driving back in the opposite direction. Mathematically we write this as

$$v(t) = t^2 - 8t + 12 < 0. \tag{4.17}$$

Likewise, between the hours of $x = 0$ and $x = 2$ and again between $x = 6$ and $x = 8$ the DISTANCE is rising so the VELOCITY is positive during those hours. Mathematically we write this as

$$v(t) = t^2 - 8t + 12 > 0. \tag{4.18}$$

The expressions 4.17 and 4.18 are referred to as **quadratic inequalities**. In both 4.17 and 4.18 it is possible to solve each inequality analytically by using factoring:

$$v(t) = t^2 - 8t + 12 = (t-2)(t-6) < 0. \tag{4.19}$$

The factored inequality $(t-2)(t-6) < 0$ in 4.19 implies that we want the product of two values $(t-2)$ and $(t-6)$ to be negative. That is we wish to determine those values of x that make

$$(t-2) > 0 \text{ and } (t-6) < 0$$

or

$$(t-2) < 0 \text{ and } (t-6) > 0.$$

Clearly, the product $(t-2)(t-6) = 0$ when $t = 2$ or $t = 6$. This partitions the set of real numbers into three sets: those numbers less than 2, those numbers between 2 and 6, and those numbers greater than 6. Any number that is chosen that is less than 2 and substituted into the product $(t-2)(t-6)$ will produce the same sign for the product $(t-2)(t-6)$. Likewise any number that is chosen between 2 and 6 and substituted into the product $(t-2)(t-6)$ will produce the same sign for the product $(t-2)(t-6)$. Finally, any number that is chosen greater than 6 and substituted into the product $(t-2)(t-6)$ will produce the same sign for the product $(t-2)(t-6)$. Therefore, we need only pick one number from each set, substitute these numbers into the product $(t-2)(t-6)$ and determine if the product is then positive or negative. We will choose numbers as easy as possible: for the set of numbers less than 2 choose $t = 0$, for the set of numbers between 2 and 6 choose $t = 3$, and for the set of numbers greater than 6 choose $t = 7$.

$$
\begin{aligned}
t &= 0; \ (0-2)(0-6) > 0 \\
t &= 3; \ (3-2)(3-6) < 0 \\
t &= 7; \ (7-2)(7-6) > 0.
\end{aligned}
$$

Therefore, the solution to 4.17 is

$$\{t \mid 2 < t < 6\}.$$

Note that this factoring process automatically solves the inequality 4.18 as well. The inequality 4.18 has solution

$$\{t \mid t < 2 \text{ or } t > 6\}.$$

Try the following exercises in class.

Exercise 92 *In this activity you are monitoring the rise and fall of several stocks over the 8 hour trading day. The expressions below refer the RATE at which each stock is changing throughout the day. Use factoring to determine when each stock is rising and when each is falling.*

1. Stock 1 Rate of Change: $f(t) = t^2 - 9t + 20$

2. Stock 2 Rate of Change: $f(t) = t^2 - 8t + 7$

3. Stock 3 Rate of Change: $f(t) = t^2 - 5t + 6$

Exercise 93 *This technique can also be used to answer additional questions about the model. Answer each of the following questions using the stock models in the above exercise.*

1. At what times during the day was Stock 1 rising at $2.00 per hour?

   2. At what times during the day was Stock 3 rising at $20.00 per hour?

   3. At what times during the day was Stock 2 falling at $9.00 per hour?

4.4.3   The Quadratic Formula and Completing the Square

Note that in each of the exercises in the previous section the expressions factored completely. This is rarely the case in real situations. The question is what do we do if faced with a quadratic model that does not factor so nicely? For example, in the previous section 4.1.2 if we use example of Stock 2, how would we determine the answer to the following question?

  At what times during the day was Stock 2 falling at $4.00 per hour?   (4.20)

   Since we are interested in when the stock is falling at $5.00 per day this implies that the equation we wish to solve is

$$t^2 - 8t + 7 = -4$$

or

$$t^2 - 8t + 11 = 0 \qquad (4.21)$$

Clearly, the quadratic in 4.21 is not factorable. So how would we find a solution? The answer is in a method called **completing the square** and a formula which it produces (one you are familiar with) called the **quadratic formula**. Here it is:

Algorithm 94 *Steps to completing the square for a quadratic equation:* $ax^2 + bx + c = 0$

1. *Divide through by the squared coefficient*

$$x^2 + \frac{b}{a}x + \frac{c}{a} = 0$$

2. *Move the resulting constant term to the other side of the equation*

$$x^2 + \frac{b}{a}x = -\frac{c}{a}$$

3. *Take half of the linear coefficient* ($\frac{b}{2a}$), *square it* ($\frac{b^2}{4a^2}$) *and add it to both sides of the equation*

$$x^2 + \frac{b}{a}x + \frac{b^2}{4a^2} = -\frac{c}{a} + \frac{b^2}{4a^2}$$

4. *Combine fractions on the right-hand side using a common denominator*

$$x^2 + \frac{b}{a}x + \frac{b^2}{4a^2} = \frac{b^2 - 4ac}{4a^2}$$

5. *Factor the left-hand side of the equation*

$$\left(x + \frac{b}{2a}\right)^2 = \frac{b^2 - 4ac}{4a^2}$$

6. *Take the square root of both sides of the equation*

$$x + \frac{b}{2a} = \frac{\pm\sqrt{b^2 - 4ac}}{2a}$$

7. *Solve for the variable x*

$$x = \frac{-b}{2a} \pm \frac{\sqrt{b^2 - 4ac}}{2a} = \frac{-b \pm \sqrt{b^2 - 4ac}}{2a}. \tag{4.22}$$

Note that the result of this algorithm is the quadratic formula 4.22. Instead of going through this process each time we can simply memorize the formula in 4.22 and use it when needed. Thus, using 4.22 in reference to solving 4.21 we have

$$a = 1; b = -8; \text{ and } c = 11.$$

Substituting these values into 4.22 we have

$$x = \frac{-(-8) \pm \sqrt{(-8)^2 - 4 \cdot 1 \cdot 11}}{2 \cdot 1} = \frac{8 \pm \sqrt{20}}{2} = 4 \pm \sqrt{5}.$$

Thus, $x \approx 6.236$ or $x \approx 1.764$. This says that approximately after the first hour and a quarter the stock is falling at \$5.00 per hour and approximately after six and a quarter hours the stock again is falling at \$5.00 per hour.

As an in-class exercise try the following.

Problem 95 *At what times during the day was Stock 1 (from section 4.1.2) rising at \$1.00 per hour?*

As a final note the quadratic formula can always be relied on to obtain an answer to a quadratic equation or inequality. In the case of a factorable quadratic, if we chose to use the quadratic formula instead of factoring the answer we obtained would be the same as if we had factored it. As a result we tend to use the quadratic formula only when necessary.

4.4.4 Problems

1. Factor each of the following quadratics.

   (a) $x^2 + 15x + 50$

   (b) $x^2 - 10x - 39$

   (c) $x^2 - 15x + 26$

   (d) $2x^2 - 5x - 3$

   (e) $64x^4 - y^4 z^8$

   (f) $64x^3 + y^6 z^3$

   (g) $x^2 + 16xy + 64y^2$

   (h) $27 - a^3$

   (i) $3x^2 + 15x - 2xy - 10y$

   (j) $10x^2 + 21xy + 9y^2$

2. You have been asked by the Department of Environmental Resources (DER) to monitor the water flow (gal. per hour) of several small streams in the Cumberland Valley that have been diverted into a local reservoir. After collecting data over the life of the storm the field teams you have set up provide you with a report detailing the following flow-rate information (flow-rate is recorded in *(gallons per hour) per hour*, the testing lasted over a continuous 8 hour period during the day):

Stream 1	$f_1(t) = t^2 - 8t + 15$ (gal. per hour) per hour
Stream 2	$f_2(t) = t^2 - 5t + 4$ (gal. per hour) per hour
Stream 3	$f_3(t) = t^2 - t - 6$ (gal. per hour) per hour
Stream 4	$f_4(t) = t - 4$ (gal. per hour) per hour
Stream 5	$f_5(t) = -2t + 6$ (gal. per hour) per hour

   (a) Determine where each creek is increasing its flow of water and decreasing its flow of water. Explain how you know this.

   (b) Sketch a graph of each flow-rate over 8 hours in the grid provided.

   (c) What is the flow-rate for each creek 3 hours into the storm?

   (d) At what time is the flow-rate of Stream one increasing at 3 (gal. per hour) per hour?

   (e) At what time is the flow-rate of Stream four increasing at 5 (gal. per hour) per hour?

   (f) At what time is the flow-rate of Stream three decreasing at 6 (gal. per hour) per hour?

3. You are monitoring the rise and fall of several stocks over the 8 hour trading day. The expressions below refer the RATE at which each stock is changing throughout the day.

Stock A	$f_A(t) = t^2 - 13t + 42$
Stock B	$f_B(t) = t^2 - 7t + 10$

(a) Determine when each stock is rising and when each is falling throughout the day.

(b) At what times during the day was Stock 4 rising at $12.00 per hour?

(c) At what times during the day was Stock 5 falling at $2.00 per hour?

(d) At what times during the day was Stock 5 falling at $1.00 per hour?

4.5 Translation of Axes

The previous sections of this chapter introduced the quadratic function model, $f(x) = ax^2 + bx + c$. The graph of the **standard quadratic function model**, $f(x) = x^2$, where $a = 1$, $b = 0$ and $c = 0$ is shown here.

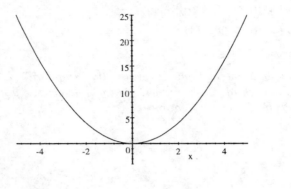

(4.23)

Standard Quadratic Graph

This section looks at some of the graphs that can be created from this model 4.23. That is, since every quadratic has the same general curve, that of a parabola, we can use this information to help make graphing quadratic functions easier. If we were graphing a linear model we would need only plot two distinct points (recall from geometry that two points determine a line). However, the more complicated a function is the more points are required to graph it accurately. This can be very time consuming so we must look for better ways to graph models without having to always rely on the inexact method of plotting points. We will use a TI-92 in this exercise to help build graphing techniques.

Answer each of the following.

Exercise 96 *Begin by plotting a few points from the **standard quadratic function model** $f(x) = x^2$ to see if you agree with the graph given above. As mentioned earlier, the curve itself is referred to as a parabola and has many of special properties and applications. One of the more common applications of the parabola is in their use as special reflecting mirrors used to enhance light sources. This is the shape used in the reflectors of flashlights and car headlights as well as telescopes to name a few.*

1. *First we will set up the graphing calculator. For the TI-92*

    (a) *Key Green Diamond (GD), Window. Window settings should be: x range: -10 to 10, scale 2; y range: -20 to 100, scale 5; $x - y$ ratio set at 1.*

    (b) *Key GD, y =. Type in the y1 = position the standard quadratic function model, $x\hat{\ }2$.*

(c) *Key GD, Graph. Does the parabola appear on the screen. If not, call the instructor over.*

2. *Next, Key GD, y =. Type in the $y2 =, y3 =, y4 =$ positions each of the following.*

   (a) $x^2 + 10$ *(x^2 + 10).*

   (b) $x^2 - 10$ *(x^2 - 10).*

   (c) $x^2 + 30$ *(x^2 + 30).*

   (d) *Explain the relationship between the "standard" parabola of $f(x) = x^2$ and each of $y2, y3$ and $y4$. Can you determine a general rule for the model $f(x) = x^2 + number$?*

3. *Next we will replace the graphs of $y2 =, y3 =, y4 =$ with new graphs for another exercise. Key GD, y =. Type in the $y2 =, y3 =, y4 =$ positions each of the following.*

   (a) $(x + 5)^2$ *((x + 5)^2).*

   (b) $(x - 4)^2$ *((x - 4)^2).*

   (c) $(x + 8)^2$ *((x + 8)^2).*

   (d) *Explain the relationship between the "standard" parabola of $f(x) = x^2$ and each of $y2, y3$ and $y4$. Can you determine a general rule for the model $f(x) = (x + number)^2$?*

Exercise 97 *Additional parabolas and their rules for graphing.*

1. *Plot the standard parabola $f(x) = x^2$ and each of the following quadratics on the graphing calculator. Explain how each differs from the standard quadratic $f(x) = x^2$.*

(a) $f(x) = 2x^2$

(b) $f(x) = 3x^2$

(c) $f(x) = \frac{x^2}{2}$ *or* $f(x) = \frac{1}{2}x^2$

(d) $f(x) = \frac{x^2}{3}$ *or* $f(x) = \frac{1}{3}x^2$

4.5.1   Problems

1. Graph each of the following using the techniques of translation of axes.

   (a) $f(x) = (x - 2)^2$

   (b) $f(x) = (x + 5)^2 - 3$

   (c) $f(x) = -(5 - x)^2 + 1$

   (d) $f(x) = 2(x - 1)^2$

   (e) $f(x) = 2(x + 2)^2 - 2$

2. Graph each of the following using the techniques of translation of axes. (Hint: use completing the square first)

   (a) $f(x) = x^2 + 4x + 6$

   (b) $f(x) = x^2 - 6x + 2$

   (c) $f(x) = x^2 - 5x + 3$

   (d) $f(x) = x^2 - 2x - 5$

   (e) $f(x) = -x^2 + 8x$

3. Give a functional expression for each of the following graphs.

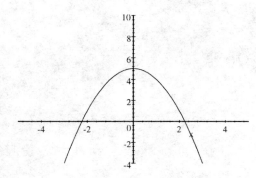

(a)

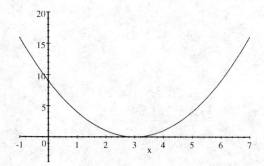

(b)

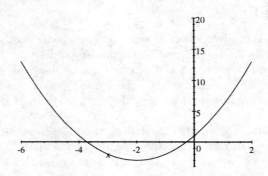

(c)

4.6   Summary

A good rule of thumb is to always keep it simple. The second most useful model is the quadratic model. As in the case of the linear model very little data is needed to develop a quadratic model. Of course, more data can only give us more confidence that the chosen model is the correct one. It is also easy to develop the geometric, analytic and numerical representations of the quadratic model.

Once again, real data seldom fits exactly into the mathematical model used to represent it. However, through regression techniques we can find the quadratic curve of best fit. The geometric curve that represents a quadratic function is called a parabola. Thus, if we plot a given set of data and the data appears to follow a parabolic path, then we can apply a quadratic model to the data. If the data appears to follow the path of a line then a linear model is used.

The quadratic model also has many practical applications from the use of parabolic mirrors in flashlights, automobile headlights and telescopes to population dynamics. We have also seen how quadratic models play an introductory role in decision making processes. This application will be developed further in future chapters.

Both the linear and quadratic models have many applications and are easily developed through geometric, analytical and numerical methods. As we continue to develop new types of models we will continue to compare the results to that of linear and quadratic models.

Chapter 5

Models Using Averages

5.1 Introduction

One of the most applicable mathematical concepts is a concept students to which are first introduced in elementary school. This is the concept of *average*. The concept of average is based upon the idea that when estimating an unknown quantity several trial measurements are taken. When several trial measurements are taken it is believed the true quantity is located centrally within the trial measurements. Averaging is the process of locating this central measurement.

In this chapter the notion of average is presented by introducing the reader to several models that apply the concept to some common problems. Since several of the examples involve multiple dimensions the chapter begins with a review of the coordinate plane. The remaining sections of the chapter discuss a variety of definitions of average as well as several applications.

5.2 The Coordinate Plane

Coordinatizing the plane is the process of identifying points in a plane using a *coordinate system* that locate points in the plane through reference axes and an origin. The most common way to coordinatize the plane is the **Rectangular Coordinate System**, also known as the **Cartesian Coordinate System**. This name refers to seventeenth century mathematician and philosopher René DesCartes, the man given credit for developing the system. His original purpose was to develop a system for drawing maps for trade ships sailing the western coasts of Europe and Africa. Today, the field of map making is referred to as **cartography** also in honor of DesCartes.

The Rectangular Coordinate System identifies every point in the plane with an **ordered pair** (x, y) in which x is the point's horizontal "distance" from a fixed origin along a **horizontal axis** (referred to as the **x-axis**) and y is the point's vertical "distance" from a fixed origin along a **vertical axis** (referred to as the **y-axis**).

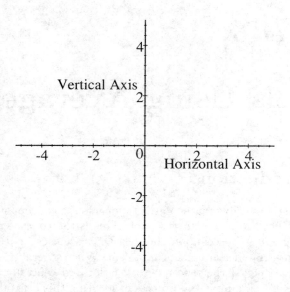

$$(5.1)$$

The Coordinate Plane

Diagram 5.1 shows a standard rectangular coordinate system with both the *horizontal axis* and the *vertical axis* labeled. Both axes have the positive and negative directions marked in the diagram, the positive x-direction is to the right and the positive y-direction is up. To identify points in the plane we simply identify the *horizontal coordinate* (x) and *vertical coordinate* (y) along each axis that corresponds to the point's position. The horizontal coordinate is recorded first and the vertical coordinate is recorded second. Note that the axes subdivide the plane into four sections. These sections are referred to as **quadrants**. The first quadrant is the quadrant in which both the x-coordinate and the y-coordinate are positive $(+, +)$. The second, third and fourth quadrants are numbered counter-clockwise from the first quadrant, respectively.

Applications that make use of a coordinate system are plentiful. Several applications are

Example 98 *road maps where a street index lists streets and their coordinates where they can be found on the map.*

Example 99 *airport control towers where air traffic controllers need to know the exact location of all planes in the immediate vicinity at all times. (they use a polar coordinate system, however, instead of a rectangular one. Why?)*

Example 100 *computer-generated images which are divided into very "small" rectangles (pixels) each of which is identified numerically by the computer so that it can "reconstruct" the picture in new locations (such as the internet).*

Use the space below to list additional applications that require the use of a coordinate system.

The rectangular coordinate system has proven very useful in previous chapters. In this chapter its applicability will be extended to averages. Our discussion begins with the "one-dimensional" standard average, referred to as the arithmetic average.

5.3 The Arithmetic Average

5.3.1 One-Dimensional Averages

The most common average formula, the one familiar to elementary students, is that of the arithmetic average.

Definition 101 *The **arithmetic average** of a set of numbers $\{x_1, x_2, ..., x_n\}$ is defined to be*

$$\frac{x_1 + x_2 + ... + x_n}{n},$$

That is, it's the sum of the numbers divided by the total number of numbers.

Exercise 102 *Find the average of the following list of numbers.*

$$\{78, 90, 86, 75, 90, 86, 88, 90\}.$$

Note that in the above exercise 102 both of the numbers 86 and 90 occur more than once. This observation will be used to introduce the concept of **weighted averages**.

Another way to view the average of the list of numbers from exercise 102 is the following: if a one-gram weight were placed at the appropriate point on a number-line for each of the numbers in the list, the average would be where a fulcrum would be placed to "balance out" the weight (see diagram 5.2).

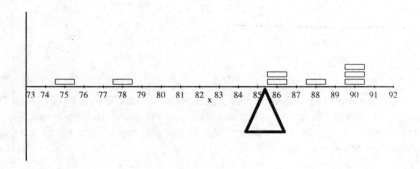

Average as Weights

$$(5.2)$$

The place of the fulcrum and the "one gram" weights in diagram 5.2 is analogous to averaging the numbers from the list

$$\{78, 90, 86, 75, 90, 86, 88, 90\}.$$

As worked out in exercise 102 this average is

$$\frac{75 + 78 + 86 + 86 + 88 + 90 + 90 + 90}{8} = 85.375 \qquad (5.3)$$

Note that in diagram 5.2 the point 90 has three weights placed on it due to the fact that 90 appears three times in the list. Likewise, the point 86 has two weights placed on it due to the fact that 86 appears twice in the list. Note also that each of the numbers 75, 78 and 88 have one weight placed on each of their positions. In that case we say that the number 90 has *weight* 3 and the number 86 has *weight* 2, while the numbers 75, 78 and 88 each have weight one. Therefore, an alternative way of stating exercise 102 is the following.

Example 103 *The list of numbers* $\{75, 78, 86, 88, 90\}$ *are given weights* $1, 1, 2, 1, 3$, *respectively and have* weighted average

$$\frac{75 \cdot 1 + 78 \cdot 1 + 86 \cdot 2 + 88 \cdot 1 + 90 \cdot 3}{1 + 1 + 2 + 1 + 3} = 85.375$$

This example leads to the following definition of a one-dimensional weighted average.

Definition 104 *If the numbers* $\{x_1, x_2, ..., x_n\}$ *are assigned the weights* $\{m_1, m_2, ..., m_n\}$, *respectively, then the list* $\{x_1, x_2, ..., x_n\}$ *has* weighted average

$$\frac{x_1 \cdot m_1 + x_2 \cdot m_2 + ... + x_n \cdot m_n}{m_1 + m_2 + ... + m_n}. \tag{5.4}$$

Clearly, this concept of a weighted average has many applications. For example, exercise 102 could represent the grades a student has received on exams over the course of a term. This is similar to the technique used to compute a students overall grade point average (GPA). The credits earned for a course is the weight assigned to the course while the grade for the course is the value assigned to that course ($A = 4$ points, $B = 3$ points,...). The student's GPA for the semester is then computed by multiplying the number of credits for a course by the numerical grade earned for the course. These products are then summed and divided by the total number of credits. Additional applications can be found in the exercises at the end of the section.

5.3.2 Two-Dimensional Averages

There are many applications in which we are interested in finding the "center" of a series of points in the coordinate plane. This section will outline the process of finding the two-dimensional weighted average of a series of points in the plane. We begin by reviewing a formula that is familiar to most algebra students. The simplest application of a two-dimensional weighted average is that of finding the point located exactly midway between two given points in the plane. This is equivalent to finding the midpoint of the line segment joining the given points (x_1, y_1) and (x_2, y_2). This process results in the midpoint formula.

Definition 105 *The **midpoint** $(\overline{x}, \overline{y})$ of two points in the rectangular coordinate* (x_1, y_1) *and* (x_2, y_2) *is*

$$(\overline{x}, \overline{y}) = \left(\frac{x_1 + x_2}{2}, \frac{y_1 + y_2}{2} \right).$$

Note that this formula appears to be an average of sorts. That is, the $x - corrdinate$ of the midpoint is the average of the $x - coordinates$ and the $y - corrdinate$ of the midpoint is the average of the $y - coordinates$. Note also that this averaging of the coordinates produces the midpoint of the line segment joining the original two points.

This concept is easily extended to finding the center of a finite cluster of points in the plane. That is, the $x - corrdinate$ $(\overline{x})$ of the center is the average of the $x - coordinates$ and the $y - corrdinate$ $(\overline{y})$ of the center is the average of the $y - coordinates$.

Exercise 106 *Find the average of the points* $(3, -1), (5, 4), (-2, 6)$ *and* $(-5, -6)$ *(see diagram 5.5).*

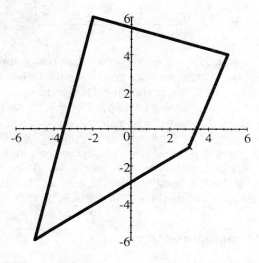

(5.5)

Averaging Four Points in the Plane

Finding the average of the four points (vertices) defining the quadrilateral in figure 5.5 is equivalent to finding the point at which a fulcrum can be placed under the quadrilateral to have it balance. That is, if we cut the quadrilateral from a piece of cardboard and marked the average of its four vertices on the cut out, we could balance the quadrilateral on the tip of a pencil by placing the pencil point at the average point. We have found the *center of mass* of the quadrilateral.

It should be clear that if we attach a one gram weight to each of the four vertices, the quadrilateral will still balance at the same point (the center of mass remains the same). What happens though if unequal masses are placed at the four vertices? As might be expected from the discussion of one-dimensional weighted averages, the center of mass will change its location and the quadrilateral will no longer balance. The two dimensional weighted average defined below is exactly the mechanism that locates the new center of mass. The two dimensional weighted average is an extension of both the two dimensional average and the one dimensional weighted average; that is, the $x-coordinate$ ($\overline{x}$) of the two dimensional weighted average is the weighted average of the x coordinates and the $y-coordinate$ ($\overline{y}$) is the weighted average of the y coordinates.

Definition 107 *The **weighted average**, $(\overline{x}, \overline{y})$ of a set of points $\{(x_1, y_1), (x_2, y_2), ..., (x_n, y_n)\}$ with weights $\{m_1, m_2, ..., m_n\}$, respectively, is*

$$(\overline{x}, \overline{y}) = \left(\frac{x_1 \cdot m_1 + x_2 \cdot m_2 + ... + x_n \cdot m_n}{m_1 + m_2 + ... + m_n}, \frac{y_1 \cdot m_1 + y_2 \cdot m_2 + ... + y_n \cdot m_n}{m_1 + m_2 + ... + m_n} \right).$$

(5.6)

Industry provides many practical examples of using weighted averages. One such example is locating the source of a spill.

Exercise 108 *An official for the Environmental Protection Agency (EPA) is investigating a toxic spill at a landfill site. A toxic substance has been illegally dumped at the site and it is the official's job to locate the source of the spill (since the substance has seeped into the soil it is undetectable by sight). Assuming that the sight is square in shape, the official superimposes a rectangular grid on a map of the site to identify specific positions at the site. The official then assigns teams of workers to collect soil samples at various locations. The soil samples will be tested for the level of toxicity. The following information is recorded.*

Example 109   *1. The coordinates of where each sample was taken. This is the set*

$$\{(-2,3),(1,1),(4,5),(-3,-2),(-4,5),(-5,-5),(5,4),(4,-2),(3,-3),(0,-4)\}$$

2. The "weights" of each coordinate will represent the level of toxicity (parts per 1000 in grams) of the soil sample. Each sample location from 1 above was found to have toxicity levels

$$\{8,23,13,5,6,2,17,14,18,3\},$$

respectively.

3. Use the information from 1 and 2 to estimate the source of the toxic spill.

4. The weighted average found in 3 depends upon the choice of sample points in 1. As a result it may differ from the true center of mass. Explain how the official might better isolate the source of the spill. That is, what would you do to increase the accuracy of your answer?

With the pattern established with both the one-dimensional and two-dimensional cases it should not be difficult to see how to extend this concept to three dimensions. In fact, the definition is analogous to both cases.

Definition 110 *The three-dimensional **weighted average**, $(\overline{x}, \overline{y}, \overline{z})$ of a set of points $\{(x_1, y_1, z_1), (x_2, y_2, z_2), ..., (x_n, y_n, z_n)\}$ with weights $\{m_1, m_2, ..., m_n\}$, respectively, is*

$$(\overline{x}, \overline{y}, \overline{z}) = \left(\frac{x_1 \cdot m_1 + x_2 \cdot m_2 + ... + x_n \cdot m_n}{m_1 + m_2 + ... + m_n}, \right.$$
$$\frac{y_1 \cdot m_1 + y_2 \cdot m_2 + ... + y_n \cdot m_n}{m_1 + m_2 + ... + m_n},$$
$$\left. \frac{z_1 \cdot m_1 + z_2 \cdot m_2 + ... + z_n \cdot m_n}{m_1 + m_2 + ... + m_n} \right). \tag{5.7}$$

Several examples of all three cases are included in the exercise set.

5.3.3 Problems

1. You are the site planning manager of a large corporation. Currently you have four warehouses located at various locations from which you distribute your products. Your department must find a site to locate a new manufacturing plant which will be shipping products to all four warehouses. After superimposing a grid on a map, you can locate the warehouses by the coordinates given in the following table. Also listed is the volume (amount of products being shipped) that each warehouse is expecting to process. Based on this information, where should the new plant be located?

Warehouse	x-coord	y-coord	Volume
Tuscon	-15	-5	7.8
Chicago	5	5	9.3
San Francisco	-24	3	10.8
Atlanta	8	-3	11.6

2. Your school has purchased a set of recycling bins. You have been assigned the task of finding the ideal spot within the school complex to place the set of recycling bins. This is to be done according to student-density population (a map of the campus would be helpful). For example, if there are residence halls on campus and residence hall A holds more students than residence hall B, then the bins should be placed proportionately closer to hall A than hall B. If the campus does not have residence halls, then "student density" may be determined by how often a classroom is used or what the capacity of that classroom is.

    (a) On your own define what is meant by "student density" (i.e., capacity of residence halls, capacity or use of classrooms, etc.) and use this definition to determine the capacity of each unit (residence hall, classroom, etc.).

    (b) Describe a procedure that will determine the best "student-density" location for the set of recycling bins.

    (c) Use the data you collected and the procedure you described to calculate where that best location is.

    (d) Draw some conclusions about your answer. That is, does it make sense? If so, explain. If not, why not and what would you do to change it?

3. Repeat exercise 2 above if two sets of recycling bins are purchased instead of one set.

4. Although exercises 2 and 3 refer to a hypothetical case, it is often the case that studies are conducted in order to convince the administration or authorities that something should be done. Use your results in problems 2

and 3 to write a report to your school administrators that convinces them that they should purchase a set of recycling bins.

5. Is your school "centrally located?" Many students do not live "on sight," that is, students commute to campus. This exercise is designed to determine if your school is in a central location for its commuters.

   (a) Conduct a survey by randomly sampling students on campus asking them what their address is. (About 20 to 30 students should be sufficient.)

   (b) Use the address data from your survey to calculate an "ideal" central location for your school.

   (c) Does the central location found in part b above agree with the school's current location? Does this answer make sense? Explain why or why not.

6. **(Internet Exercise)** A new community resource center is to built to serve a 400 square mile area.

   (a) Search the internet for a map that outlines a 400 square mile area.

   (b) Identify population centers in this area and assign each population center a population density.

   (c) Use the data in part b. to identify a central location for the new resource center.

   (d) Draw some conclusions about your answer. That is, does it make sense? If so, explain. If not, why not and what would you do to change it?

7. **(Internet Exercise)** Repeat exercise 6 using a **different** location that covers a 900 square mile area.

8. **(Internet Exercise)** A new fire house is to be built for a volunteer fire company that services a region covering a 400 square mile area. There are two ways in which the new location for the fire house can be identified: i. by the location of the population it will serve, or ii. by the location of where the fire house volunteers live (note that a volunteer fire-person must live within the area the fire company serves).

   (a) Search the internet for a map that outlines a 400 square mile area.

   (b) Identify population centers in this area and assign each population center a population density.

   (c) Use the data in part b. to identify a central location for the new fire house.

(d) If the locations of where the volunteer fire-persons live were known, would you expect the answer obtained using this information to be the same or different than the answer found in part c? If so, explain. If not, why not and what would you do to change it?

9. **(Internet Exercise)** Repeat exercise 8 using a **different** location that covers a 900 square mile area.

5.4   The Geometric and Harmonic Averages

The arithmetic average is not the only way in which data points can be averaged. There are several other types of averages, the most common of which are the geometric and harmonic averages. This section looks at applications of these averages.

5.4.1   The Geometric Average

Although not as well known as the arithmetic average, the geometric average is just as useful. It occurs in many statistical and financial applications. For example, the geometric average is used in finance to calculate average growth rate given compound interest with variable rates. It is also associated with the concept of ratio and proportion. That is, if a, b and c are real numbers such that

$$\frac{a}{b} = \frac{c}{a},$$

then

$$a^2 = bc$$

or

$$a = \sqrt{bc}.$$

We say that a is the *geometric mean* of b and c. This concept can be expanded to define the geometric mean for a finite list of numbers.

Definition 111 *Let* $\{x_1, x_2, ..., x_n\}$ *be a set of numbers. Then the geometric mean G of* $\{x_1, x_2, ..., x_n\}$ *is defined to be*

$$G = \sqrt{x_1 \times x_2 \times ... \times x_n}.$$

5.4.2   The Harmonic Average

The harmonic mean is the reciprocal of the arithmetic mean of reciprocals. That is,

Definition 112 *Let* $\{x_1, x_2, ..., x_n\}$ *be a set of numbers. Then the harmonic mean H of* $\{x_1, x_2, ..., x_n\}$ *is defined to be*

$$\frac{1}{H} = \frac{1}{n}\left(\frac{1}{x_1} + \frac{1}{x_2} + \cdots + \frac{1}{x_n}\right).$$

5.4.3 Problems

1. Determine the geometric average for each set of data points.

    (a) $\{1, 2, 3, 4, 5, 6, 7, 8, 9, 10\}$

    (b) $\{2, 4, 6, 8, 10, 12\}$

    (c) $\{-1, 2, -3, 4, -5, 6, -7, 8, -9, 10\}$

    (d) $\{12, 32, -15, 24, 40, -28, 5, 20\}$

    (e) $\{1, \frac{1}{2}, \frac{1}{3}, \frac{1}{4}, \frac{1}{5}, \frac{1}{6}, \frac{1}{7}, \frac{1}{8}, \frac{1}{9}, \frac{1}{10}\}$

2. Determine the harmonic average for each set of data points.

    (a) $\{1, 2, 3, 4, 5, 6, 7, 8, 9, 10\}$

    (b) $\{2, 4, 6, 8, 10, 12\}$

    (c) $\{-1, 2, -3, 4, -5, 6, -7, 8, -9, 10\}$

    (d) $\{12, 32, -15, 24, 40, -28, 5, 20\}$

    (e) $\{1, \frac{1}{2}, \frac{1}{3}, \frac{1}{4}, \frac{1}{5}, \frac{1}{6}, \frac{1}{7}, \frac{1}{8}, \frac{1}{9}, \frac{1}{10}\}$

5.4.4   Linear Regression

Regression is the process of fitting a curve to data. Many fields of study require collecting and studying data. However, in the process of collecting data from an experiment there are many reasons why the data does not fall exactly on a curve. In such instances we must find a curve that "fits" the data as well as possible (one that passes through the data in a general sense). There are primarily two aspects of regression that we will be concerned with here:

1. We must first determine what type of curve (linear, quadratic, exponential, etc.) best resembles the general path the data is taking. As of this stage we have but one model at our disposal, the linear model. In this section we will discuss ways of developing a linear model from given data: **linear regression**.

2. Once the type of curve is determined then we must find an analytic expression for the curve that we can use to analyze the data.

To develop a linear model from a given set of data means that we must determine an equation $y = mx + b$ that best fits the data. That is, we must find two values: the slope of the line, m and the y-intercept, b. One approach is to plot the data on a grid and then, to the best of our ability, sketch a line through the data that we think best follows the path of the data. Developing a linear model this way does is not very good because it does not give us any way of knowing how accurate our attempt is. Another way would be to use the **method of least squares**. For the moment we will give the formulas for determining the slope and the y-intercept. An explanation will follow later.

Algorithm 113 *Linear Regression - Finding the line ($y = mx + b$) of best fit for a given set of data $\{(x_1, y_1), (x_2, y_2), \ldots, (x_n, y_n)\}$. The calculations for m and b are based on the following formulas for slope and y-intercept:*

1. $m = \frac{n - (x_1 y_1 + x_2 y_2 + \cdots + x_n y_n)(x_1 + x_2 + \cdots + x_n)(y_1 + y_2 + \cdots + y_n)}{n(x_1^2 + x_2^2 + \cdots + x_n^2) - (x_1 + x_2 + \cdots + x_n)^2}$

2. $b = \frac{(y_1 + y_2 + \cdots + y_n)(x_1^2 + x_2^2 + \cdots + x_n^2) - (x_1 + x_2 + \cdots + x_n)(x_1 y_1 + x_2 y_2 + \cdots + x_n y_n)}{n(x_1^2 + x_2^2 + \cdots + x_n^2) - (x_1 + x_2 + \cdots + x_n)^2}$

The accuracy of the line calculated by the formulas in 113 depends on the degree of scatter in your data. The less linear the data, the less accurate the linear model. Here is a simple exercise:

Exercise 114 *The university has keep data over the last 20 years as to the amount of waste that has been generated by the university and surrounding community. This data is in the chart below:*

Year	1982	1987	1992	1997
Waste (in tons)	623	628	634	642

(5.8)

1. Plot this data on the grid below. Label the axes and points clearly.

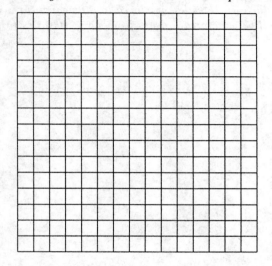

2. Plug the data from 5.8 into the formulas 113 to find the slope and y-intercept for the line of best fit.

3. Sketch the line that you found in part 2. in the grid above.

4. *We can use this regression equation that we found to make predictions as to how much waste will be produced in the near future. Estimate how much waste will be generated 5 years from now; 10 years from now.*

5.4.5 Problems

1. The table 5.9 below shows the census population of the United States from 1950 to 1990.

	Population
Year	Num.
(x=1)1950	151.3
(x=2)1960	179.3
(x=3)1970	203.3
(x=4)1980	226.5
(x=5)1990	248.7

(5.9)

(a) Plot this data on a grid. Label the axes and points clearly.

(b) Plug the data from 5.9 into the formulas 113 to find the slope and y-intercept for the line of best fit.

(c) Sketch the line that you found in part 2. in the grid from part a.

(d) Use the line of best fit found in part b to estimate the US population in the years 2000, 2010 and 2020. Explain which of these estimates in which you have the most confidence and why.

2. Vacation travel has always been an essential part of the American Dream. Each year millions of Americans plan vacations. One mode of transportation that has been growing in popularity in recent years is air travel. However, with the increased popularity of air travel comes an increase in the headaches and problems of air travel as well. One indication of this is the increase since 1990 of the number of passengers bumped from U.S. based flights. The chart below represents the number of passengers bumped from U.S. flights (in millions) since 1990 (Source: Newsweek)

	Pass. Bumped
Year	Numerical
$(x = 0)1991$	0.65
$(x = 2)1993$	0.7
$(x = 4)1995$	0.85
$(x = 6)1997$	1.05

(5.10)

(a) Plot this data on a grid. Label the axes and points clearly.

(b) Plug the data from 5.10 into the formulas 113 to find the slope and y-intercept for the line of best fit.

(c) Sketch the line that you found in part 2. in the grid from part a.

(d) Use the line of best fit found in part b to estimate the number of airlines passengers bumped in the years 2000, 2005 and 2010. Explain which of these estimates in which you have the most confidence and why.

3. Global warming is of great concern to many scientists around the world. Although there is no doubt that temperatures are on the rise, there is much disagreement as to the causes of those higher temperatures. It is the job of every good scientist to remain objective when studying data, shunning any preconceived notions or popular ideas for what is the truest nature of the observations. The following data can be extrapolated from the website *http://www.nsc.org/ehc/guidebks*. This site is one of many that discusses the rate at which major greenhouse gases are increasing in the earth's atmosphere. These gases include carbon dioxide (CO_2, from fossil fuel combustion & deforestation), methane (CH_4, from fossil fuel production, rice fields, cattle, landfills), nitrous oxide (N_2O, from nitrogenous fertilizers, deforestation, biomass burning, refrigerants, foams), and CFC-12 (from aerosol sprays, refrigerants, foams). The chart below shows the increase in the amount of methane in parts per million (ppm) in the atmosphere since the early part of this century.

	Methane
Year	Numerical
$(x=0)$1930	1.10
$(x=2)$1950	1.21
$(x=4)$1970	1.40
$(x=6)$1990	1.72

(5.11)

(a) Plot this data on a grid. Label the axes and points clearly.

(b) Plug the data from 5.11 into the formulas 113 to find the slope and y-intercept for the line of best fit.

(c) Sketch the line that you found in part 2. in the grid from part a.

(d) Use the line of best fit found in part b to estimate the number amount of methane in parts per million (ppm) in the atmosphere in the years 2000, 2010 and 2020. Explain which of these estimates in which you have the most confidence and why.

5.5 Summary

The concept of **average** is one of those mathematical concepts that was introduced at a very early mathematical age. Every school child is capable of computing a simple average of a list of data. Yet this simple concept has proven quite applicable in everyday situations. Once again, advanced mathematical concepts are not necessary in order to determine some very practical applications of mathematics.

Increasing the dimensions of the situation does not change the way in which averages are computed. If a "two-dimensional" average is required, then the averaging computations are simply completed twice, once for the horizontal component and once for the vertical component. Likewise, adding weights to the data points does not change the computational process either. Simply multiply the appropriate coordinate by the appropriate weight, sum each of the resulting products, then divide this sum by the total weight.

As this chapter demonstrated, there are many practical applications of two-dimensional weighted averages. One of the most reasonable applications is the central location of goods and services. Many companies, governments and administrations who wish to provide a service to their communities do so by determining the most reasonable location for those services. This is accomplished through the process of weighted averages. In addition, the concept of weighted averages can be used in conjunction with other mathematical concepts, such a ratio and proportion, in order to provide the user with an even more detailed analysis of the situation.

Chapter 6

Models Using Polynomials

6.1 Introduction

Polynomial models are one of the most basic functions in mathematics. They are also one of the most applicable models as well. Applications of polynomials can be found in many scientific fields (Physics, Chemistry, Astronomy, Biology and Geology, etc.), business fields (Economics, Finance, Marketing, etc.) as well as decision science, probability and many others. Basically, polynomial functions are constructed using nothing more than addition and multiplication of powers of the independent variable.

Definition 115 *A polynomial function $p(x)$ of degree n is defined by*

$$p(x) = a_0 + a_1 x + a_2 x^2 + \cdots + a_n x^n$$

Earlier chapters discussed the standard linear and quadratic polynomial models. This chapter will focus on a few of these many examples to give the reader a clear sense of how versatile these simple functions are. We begin by reviewing some basic properties of exponents and some commonly used mathematical notation.

6.2 Simplifying Expressions

Mathematical notation was not invented overnight. Mathematicians have sometimes searched for centuries for the right notation to convey a mathematical idea. It is critical not only to the understanding of mathematics but also to be able to do mathematics efficiently so that mathematical ideas can be conveyed as simply and clearly as possible. This is not easy. It requires many years and the efforts of many mathematicians before exactly the right notation is found. Don't panic! We are not going to review the history of all mathematical notation. Our focus will be narrowed to that notation specific to exponents.

175

Let's start with the relationship between addition and multiplication. The expression

$$5 \cdot 6 = 6 + 6 + 6 + 6 + 6$$

implies that "5 times 6" is interpreted to mean the repetitive process of adding 6 to itself five times. We can easily generalize this to the field of algebra where

$$
\begin{aligned}
5x + 7x &= (x + x + x + x + x) + (x + x + x + x + x + x + x) \quad (6.1) \\
&= (5 + 7)x = 12x.
\end{aligned}
$$

A similar relationship is held between multiplication and exponents. The expression

$$4^5$$

implies that "4 raised to the fifth power" is interpreted to mean the repetitive process of multiplying 4 to itself five times. Algebraically, we have

$$x^k = x \cdot x \cdots x \ (\text{k times}). \qquad (6.2)$$

That is, exponential is an abbreviated way of representing a repetitive multiplication process. The x is referred to as the **base** of the term and the power k is referred to as the **exponent** of the term. With this we are now ready to review some properties of exponents.

6.2.1   Properties of Exponents

The notation defined in 6.2 suggests a way for simplifying certain exponential expressions that are multiplied together. For example, the expression

$$x^3 \cdot x^5$$

is interpreted to mean

$$
\begin{aligned}
x^3 \cdot x^5 &= (x \cdot x \cdot x)(x \cdot x \cdot x \cdot x \cdot x) \\
&= x \cdot x \cdot x \cdot x \cdot x \cdot x \cdot x \cdot x \\
&= x^8.
\end{aligned}
$$

Since multiplication is the only operation used in this notation the process can be easily described as "the product of x with itself three times multiplied by the product of x with itself five times is the product of x with itself eight times." If x is any real number and m and n are any positive integers, then this multiplication process can be summarized in the following axiom.

Axiom 116 $x^m \cdot x^n = x^{m+n}$: *When multiplying two expressions with the same base exponents are added together.*

Several other "properties" of exponents can be developed using this reasoning. Again let x be a real number and m and n positive integers.

Axiom 117 $(x^m)^n = x^{mn}$: *When raising an exponential expression to another power the exponents are multiplied.*

Axiom 118 $\frac{x^m}{x^n} = x^{m-n}$ *(m > n): When dividing two expressions with the same base exponents are subtracted.*

Axiom 119 $\frac{x^m}{x^n} = \frac{1}{x^{n-m}}$ *(m < n): When dividing two expressions with the same base exponents are subtracted.*

The previous two axioms can be used to prove another well known property of exponents.

Axiom 120 $x^0 = 1$: *This is proven by* $x^0 = x^{m-m} = \frac{x^m}{x^m} = 1$.

Note that the exponential axioms are rules that are used when the only operation that appears in the expression is that of multiplication. When an algebraic expression involves both addition and multiplication then another set of axioms are required. These axioms will be discussed later in the section. The axioms 116, 117, 118, 119, and 120 above can be expanded to include an explanation of negative exponents. For example, 119 implies that $\frac{x^4}{x^7} = \frac{1}{x^3}$. However, 118 implies that this could also be written as $\frac{x^4}{x^7} = x^{4-7} = x^{-3}$. That is,

$$\frac{x^m}{x^n} = x^{m-n} = \frac{1}{x^{n-m}}.$$

Hence,

Definition 121 $x^{-m} = \frac{1}{x^m}$: *Negative exponents represent the reciprocal expression.*

This extension of axioms 116, 117, 118, 119, and 120 can be used to develop mathematical definitions of fractional exponents as well. For example, the expression $x^{\frac{1}{2}}$ should be defined in such a way so that it agrees with all the previous mathematical properties and definitions of exponents. That is, if we calculate $x^{\frac{1}{2}} \cdot x^{\frac{1}{2}}$ then the answer should agree with the first axiom stated above. Hence,

$$x^{\frac{1}{2}} \cdot x^{\frac{1}{2}} = x^{\frac{1}{2}+\frac{1}{2}} = x^1 = x.$$

This implies that $x^{\frac{1}{2}} = \sqrt{x}$. Therefore,

Definition 122 $x^{\frac{1}{m}} = \sqrt[m]{x}$: *An exponent of the form $\frac{1}{m}$ represents the mth root of the base.*

Definition 122and axiom 117 implies a more general rule for fractional exponents stated here as

Axiom 123 $x^{\frac{n}{m}} = \left(x^{\frac{1}{m}}\right)^n = \sqrt[m]{x^n}$.

Note that exponential properties are mathematical identities. That is, they work "both ways." Mathematical expressions are manipulated for the purpose of simplifying an expression to make it easier to solve the problem at hand. Thus, exponential expressions are often used in reverse when simplifying expressions. Let's work out some examples in class before these properties are applied to situations.

Example 124 *Simplify* $\left(3x^2y^3\right)^3$

Solution 125 $\left(3x^2y^3\right)^3 = 3^3\left(x^2\right)^3\left(y^3\right)^3 = 27x^6y^9$

Example 126 *Simplify* $\left(2x^{-4}y^3\right)\left(5x^6y^{-2}\right)$

Solution 127 $\left(2x^{-4}y^3\right)\left(5x^6y^{-2}\right) = 2\cdot 5\left(x^{-4}\right)\left(x^6\right)\left(y^3\right)\left(y^{-2}\right) = 10x^2y$

Example 128 *Simplify* $\frac{\left(3x^2y^3\right)^2}{\left(6x^{-1}y^2\right)}$

Solution 129 $\frac{\left(3x^2y^3\right)^2}{6x^{-1}y^2} = \frac{9x^4y^6}{6x^{-1}y^2} = \frac{3x^5y^4}{2}$

6.2.2   Combining Like Terms

As mentioned above, the axioms 116, 117, 118, 119, 120 and 123 are rules that govern the behavior of exponents when the only operation involved in the expression is multiplication. If the expression also involves the operation of addition then new properties must be employed.

Exercise 130 *Explain why* $x^3 + x^4$ *CANNOT be simplified to* x^7.

Exercise 131 *Explain why $\frac{x^4}{x^3+y^4}$ CANNOT be simplified to $\frac{x}{y^4}$.*

The previous exercises demonstrate that when both addition and multiplication are present in the expression the previous axioms do not apply. However, there are some properties that will be helpful.

Example 132 *Simplify the expression*

$$4x^3 + 7x^3.$$

This will be accomplished by using the distributive law described above in 6.1. Therefore,

$$
\begin{aligned}
4x^3 + 7x^3 &= (x^3 + x^3 + x^3 + x^3) + (x^3 + x^3 + x^3 + x^3 + x^3 + x^3 + x^3) \\
&= (4+7)x^3 = 11x^3
\end{aligned}
$$

That is, when four x^3's are added to seven x^3's the result is eleven x^3's.

The expression $4x^3 + 7x^3$ can be simplified because both $4x^3$ and $7x^3$ involve the expression x^3. Both $4x^3$ and $7x^3$ are referred to as the *terms* of the expression.

Definition 133 *An algebraic **term** $ax_1^{n_1} x_2^{n_2} \cdots x_k^{n_k}$ is a product of a nonzero number, a, and one or more variables x_1, x_2,..., x_k, raised to the respective powers n_1, n_2, ..., n_k. The constant a is called the **coefficient** of the term.*

Definition 134 *Two terms $ax_1^{n_1} x_2^{n_2} \cdots x_k^{n_k}$ and $bx_1^{m_1} x_2^{m_2} \cdots x_t^{m_t}$ are said to be **like terms** if $k = t$ and $n_1 = m_1$, $n_2 = m_2$, ..., $n_k = m_t$. That is, if all the variables are the same and their corresponding exponents are the same.*

Example 135 *Some examples of like terms.*

1. *$5x^4y^3$ and $8x^4y^3$ are like terms.*

2. *$-3a^6$ and $2a^6$ are like terms.*

3. *$\frac{2}{3}x^5y^3z$ and $-7x^5y^3z$ are like terms.*

These definitions provide a way to generalize the property outlined in exercise 132.

Axiom 136 *If* $ax_1^{n_1} x_2^{n_2} \cdots x_k^{n_k}$ *and* $bx_1^{n_1} x_2^{n_2} \cdots x_k^{n_k}$ *are like terms, then*

$$ax_1^{n_1} x_2^{n_2} \cdots x_k^{n_k} + bx_1^{n_1} x_2^{n_2} \cdots x_k^{n_k} = (a+b)x_1^{n_1} x_2^{n_2} \cdots x_k^{n_k}.$$

That is, when adding like terms together simply add the coefficients of each term (leaving the variables and their exponents the same).

Axiom 136 as well as the properties of exponents provide the means for simplifying algebraic expressions. Simplifying expressions will be important to the applications discussed in this chapter. Therefore, some problems on simplifying algebraic expressions are provided at the end of this section. Completing these exercises will provide the necessary review before studying the applications of polynomials in the later sections of this chapter.

6.2.3 Problems

1. Identify which axioms and properties are being used to simplify each expression.

   (a) $(7a^4)(-4a^5) = -28a^9$

   (b) $5x^3y - 7x^3y = -2x^3y$

   (c) $3xy^2(4xy + 5xy) = 27x^2y^3$

   (d) $a^3b^2(2ab + 5a^2b) = 2a^4b^3 + 5a^5b^3$

   (e) $\frac{15x^4y^2z}{5xy^3} = \frac{3x^3z}{y}$

   (f) $\frac{5a^3 + 2a^2}{3a^2} = \frac{5a+2}{3}$

   (g) $\left(\frac{34x^3y}{15xy^6}\right)^0 = 1$

   (h) $7a^3bc^2 - 5ab - 3a^3bc^2 + 4ab = 4a^3bc^2 - ab$

2. Use the properties of exponents to simplify each of the following expressions.

   (a) $(5x^4y^3)(3x^5y^3)$

   (b) $(4x^4y^3)(-12x^{-4}y^3)$

   (c) $(-15a^{-1}b^2c^3)(2a^4c^{-2})$

   (d) $(-6a^{11})(-5a^5)$

   (e) $\frac{24x^2y^2z^2}{15xyz}$

   (f) $\frac{21x^6y^2}{-14xy^5}$

   (g) $\frac{35a^{-2}b^{-1}}{15a^{-3}b^5}$

   (h) $\frac{(2x^2y^{-2}z^2)(-3x^{-3}y)}{(15xyz)(4x^{-2}y^{-3})}$

3. Use the properties of exponents to simplify each of the following expressions.

   (a) $(5x^2y)^3(x^2y^3)$

   (b) $(-2a^2b^2)^3(3ab^2)^2$

   (c) $(7x^2y)(5x^{-2}y^{-3})^2$

   (d) $(x^{\frac{2}{3}}y^{-\frac{3}{5}})(x^{\frac{1}{2}}y^{\frac{3}{5}})$

   (e) $(x^{\frac{1}{2}}y^{-\frac{3}{5}})^{\frac{1}{3}}$

   (f) $\frac{32x^{\frac{1}{3}}y^2}{-16xy^{\frac{2}{3}}}$

4. Use the properties of exponents and of combining like terms to simplify each of the following expressions.

(a) $5x^2y + 4x^2y$

(b) $7x^4y^5 + 9x^4y^5 - 3x^4y^5$

(c) $15a^3b^3 - 9a^3b^3 + 4a^3b^2$

(d) $2xy(3x + 2y)$

(c) $-4a^2b^2(2a^3 - 5ab^2)$

(f) $3xy^3(x^2y^2 - 2xy^3 + 2x^3y)$

6.3   Polynomial Models

Earlier chapters studied two special cases of polynomial models: the linear function model,

$$l(x) = ax + b,$$

and the quadratic function model,

$$q(x) = ax^2 + bx + c,$$

where $a \neq 0$. Both of these models proved to be very useful. However, there are many applications that require something more. This chapter will consider additional polynomial models of the form

$$p(x) = a_n x^n + a_{n-1} x^{n-1} + \cdots + a_1 x + a_0, \; n > 1,$$

as well as some very interesting and unusual applications of these polynomials. What follows are more examples of how polynomials are used to model certain information. These support the earlier topics and offer a more precise explanation of how the model works.

6.3.1   A Model Involving Dice

Polynomials can be used to model all sorts of experiments. Here is an exercise involving dice to demonstrate just how polynomials can be used to give us information about a situation.

Exercise 137 *Many games involve tossing two dice in order to move pieces around the board. Tossing two dice can be used as a simple example of how polynomials model experiments and situations. Color-coding the two dice, say a red die and a green die, will simplify things. The tree diagram 6.3 below demonstrates the possible combinations of numbers and their sums.*

RED DIE	GREEN DIE	SUM of DICE
	Green 1	= 2
Red 1	Green 2	= 3
	Green 3	= 4
	Green 4	= 5
	Green 5	= 6
	Green 6	= 7
	Green 1	= 3
Red 2	Green 2	= 4
	Green 3	= 5
	Green 4	= 6
	Green 5	= 7
	Green 6	= 8
	Green 1	= 4
Red 3	Green 2	= 5
	Green 3	= 6
	Green 4	= 7
	Green 5	= 8
	Green 6	= 9
	Green 1	= 5
Red 4	Green 2	= 6
	Green 3	= 7
	Green 4	= 8
	Green 5	= 9
	Green 6	= 10
	Green 1	= 6
Red 5	Green 2	= 7
	Green 3	= 8
	Green 4	= 9
	Green 5	= 10
	Green 6	= 11
	Green 1	= 7
Red 6	Green 2	= 8
	Green 3	= 9
	Green 4	= 10
	Green 5	= 11
	Green 6	= 12

$$(6.3)$$

1. In most board games what is the most common way of combining the numbers that appear on the top faces after the dice are tossed?

2. What is recorded if the toss is red 5, green 3?_____Evaluate $x^5 \cdot x^3 = $_____

3. What is recorded if the toss is red 1, green 4?_____Evaluate $x \cdot x^4 = $_____

4. Next to each sum in the tree 6.3 above write a term $x^a \cdot x^b = x^{a+b}$ that corresponds to that sum combination. Explain which exponential property demonstrates the relationship between the dice sum.

5. An effective mathematical modeling technique is to associate a mathematical process or concept with the situation at hand. In this case axiom 116 is associated with adding the numbers on the dice. This observation can be used to write a polynomial product that represents tossing two dice. If the exponents of the terms of a polynomial are thought of as the numbers on the face of a die then the polynomial

$$x + x^2 + x^3 + x^4 + x^5 + x^6$$

represents a single cubical die. Therefore, the polynomial function

$$p(x) = (x + x^2 + x^3 + x^4 + x^5 + x^6)^2 \qquad (6.4)$$

represents two dice.

6. Multiply the product 6.4 out. Explain what the coefficients and the exponents of the terms in the product represent in terms of our experiment.

Using the properties of exponents the function 6.4 simulates the sums obtained when two dice are tossed. The function 6.4 is a coarse but effective representation of the experiment of tossing two dice and recording the sum. However, as is the case with all models some information is retained and some information is lost.

Exercise 138 *The model*

$$
\begin{aligned}
p(x) &= (x + x^2 + x^3 + x^4 + x^5 + x^6)^2 \\
&= x^2 + 2x^3 + 3x^4 + 4x^5 + 5x^6 + 6x^7 + 5x^8 + 4x^9 + 3x^{10} + 2x^{11} + x^{12}
\end{aligned}
$$

simulates the tossing of two dice.

1. *List some information that is retained by the model.*

2. *List some information that is not retained by the model.*

The main relationship behind this type of model is that the exponents in the product represent the quantity in question, in this case the numbers or sum on the dice, and each coefficient in the product represents how many ways that quantity can be obtained in the experiment. This relationship between exponential properties and ways of obtaining quantities can be exploited in many ways and will be exploited in subsequent sections. Before other applications of this concept are considered, however, additional generalizations will be conducted on the dice experiment. These generalizations will be useful to future applications.

As was noted in exercise 138 some information is lost in the model. For example, the term $5x^4$ states that there are five ways to obtain a sum of four with the dice. It does not say specifically what those five ways are. However, the function model 6.4 and 138 can be altered to provide some additional information. For example, if the function 6.4

$$p(x) = (x + x^2 + x^3 + x^4 + x^5 + x^6)^2$$

is slightly changed to the function

$$p(x, y) = (x + y^2 + x^3 + x^4 + x^5 + x^6)^2 \qquad (6.5)$$

then when this function 6.5 is multiplied we have

$$
\begin{aligned}
p(x, y) \;=\; & x^2 + 2x^4 + 2x^5 + 3x^6 + 2xy^2 + 2y^2x^3 \qquad (6.6)\\
& + 2y^2x^4 + 2y^2x^5 + 2y^2x^6 + y^4 + 4x^7 \\
& + 3x^8 + 4x^9 + 3x^{10} + 2x^{11} + x^{12}.
\end{aligned}
$$

Note that whenever "y^2" appears in the product of 6.6 that implies that a "two" has been rolled on one of the dice. Thus the term $2y^2x^5$ implies that there are two ways in which to role a "two" and a "five" on the two dice. There is one "y^4" term in 6.6 as well. Obviously, this implies that there is one way to role two "twos" on the dice.

Since the y variable is distinct from the x variable, adding the y variable to the model function 6.5 is like adding a "trace" to the problem. Once the polynomial function is expanded the y variable remains "in sight" of the reader so that every choice of a "two" on the dice is highlighted in the product. This two-variable model provides much more information, specifically, those tosses that contain a "two." Of course, the added information comes at a price. The models 6.5 and 6.6 are more complex than the original model 6.4. If more information is required, say identifying those choices with a "five" on one of the dice, then a third variable could be added to the model 6.5. This would increase the complexity even more. Therefore, it is better to run the experiment twice with two different models, one that traces "twos" and one that traces "fives."

Additional variations on this experiment will be explored in the exercises. The next several sections will take advantage of these techniques by applying them to other situations.

6.3.2   Polynomial Models Applied to Decision Making

People are often faced with making choices when selecting certain items. For example, students are required to register for courses each semester. A quick review of the schedule of available courses for the next semester reveals that there are eight (8) courses (one for each scheduled period of the day) available to the student. None of these courses conflict with each other. They are 2 English courses, 3 math courses, a biology course, and 2 History courses. The polynomial function

$$
\begin{aligned}
p(x) &= (1 + x + x^2)(1 + x + x^2 + x^3)(1 + x)(1 + x + x^2) \qquad (6.7)\\
&= (1 + x)(1 + x + x^2 + x^3)(1 + x + x^2)^2\\
&= 1 + 4x + 9x^2 + 14x^3 + 16x^4 + 14x^5 + 9x^6 + 4x^7 + x^8
\end{aligned}
$$

represents the experiment of selecting several courses from the available list of eight courses above. The tree model will not be included in this example since it would be rather large.

Exercise 139 *A student has room on his/her schedule for four of the available courses.*

   1. *Explain which part of the polynomial above represents choosing four courses.*

   2. *How many ways are there of choosing five courses? Explain.*

3. *Each factor in the model 6.7 above does not distinguish between the courses. For example, the term $16x^4$ implies that there are 16 ways to take 4 courses but it does not say what those 4 courses are. Write a polynomial function model that identifies each of the following. Support your answer in each case with concise mathematical reasoning.*

   (a) *Identify the ways in which exactly one math course can be taken next semester.*

   (b) *Identify the ways in which exactly two English courses can be taken next semester.*

   (c) *Identify the ways in which exactly one math course and no biology course can be taken next semester.*

(d) *If a student must take at least one of each math, English, biology and history course write a polynomial function that represents this situation.*

4. *Consider the four-variable polynomial function*

$$p(x, y, z, w) = (1 + x + x^2)(1 + y + y^2 + y^3)(1 + z)(1 + w + w^2).$$

Describe some advantages and disadvantages to using this four-variable model to represent the scheduling exercise.

6.3.3   Polynomial Models Applied to Storage

Another example of using polynomial models to highlight available options is storage problems.

Exercise 140 *A chemical storage facility has two "smaller" storage sites, Site One and Site Two (see diagram 6.8). Site One has four storages cells and Site Two has six storage cells. Due to government safety regulations chemicals cannot be stored simultaneously in adjacent cells, that is, cells that share a common wall. It is possible, however, to store chemicals in simultaneously in diagonal cells.*

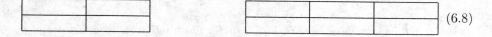 (6.8)

1. *Fill in the table below with the number of ways that the quantity of chemicals can be stored at each site.*

Number of Chemicals	Ways to Store at Site One	Ways to Store at Site Two
0		
1		
2		
3	\\\\\\\\\\\\\\\\\\\\\\\	
Polynomial		

(6.9)

2. *In the last row of table 6.9 write a polynomial under the appropriate column that represents the choices available for each site. In the space below write a polynomial product function that represents all the choices of storing chemicals at both Site One and Site Two.*

3. *Explain how many ways that five chemicals can be stored at the facility. three chemicals.*

4. *Write a two-variable polynomial that identifies the choices for storing two chemicals at Site Two.*

There are many practical examples of categorizing and storage. Additional Examples of Categorizing are

1. Police and safety agencies are interested in knowing when most automobile accidents occur. The number of accidents can be categorized according to the days of the week.

2. Universities and colleges use storage and categorizing models to help house students in residence halls on campus.

3. Scientists that study cosmic-ray experiments use storage and categorizing models: Particles (objects) reaching a Geiger Counter (cells are the counters).

4. Book publishing editors use storage and categorizing models to help determine the number of errors per page in a text to be published.

5. Biologists use storage and categorizing models to represent irradiation: Light Particles hitting the Cells of the Retina.

6. Computer Scientists use storage and categorizing models to partition the available storage cells.

Many of these examples as well as many others will be presented in the problem section.

6.3.4 Problems

1. The following questions refer to the polynomial 6.7 in section 6.3.2.

   (a) How many ways are there of choosing four courses where at least one of the courses is a History course?(Hint: this requires more than just one term)

   (b) How many ways are there of taking one math and one biology course?

   (c) How many ways are there of taking one of each type of course (math, English, history and biology)?

2. A square field (see diagram 6.10) is to be turned into an orchard. So that the trees will have sufficient room to grow the field is to be divided into 16 smaller sections and no more than two trees will be planted in each section.

   (6.10)

   (a) Write a polynomial that represents the possible ways of planting trees in one section.

   (b) Write a polynomial that represents the possible ways of planting trees in the field.

   (c) Use your answer in part b to determine the number of ways that

       i. 10 trees can be planted in the field.
       ii. 16 trees can be planted in the field.

   (d) Write a polynomial that represents the case where each section must have at least one tree in it.

3. A college residence hall has 20 rooms. Eight larger rooms have a capacity of three students each while the remaining twelve rooms have a capacity of two students.

   (a) What is the capacity of the residence hall?

   (b) Write a polynomial that represents the possible capacity choices for the eight larger rooms.

   (c) Write a polynomial that represents the possible capacity choices for the remaining twelve rooms.

   (d) Write a product of polynomials that represents the possible capacity choices for the residence hall.

   (e) Use your answer in part d to determine the number of ways that the college could house

    i. 40 students.

    ii. 30 students.

    iii. 48 students.

4. A Sampling Survey: Today there are many studies conducted by all sorts of industries. Politicians want to know where they stand with the electorate, industry wants to know if their new product will be a best seller, medical experts want to know the results of a new treatment, etc. Statisticians are asked to take a cross-section of the population in order to insure truer, less biased results. Suppose we have a group of 3 teachers, 4 lawyers, 3 doctors, 5 engineers and 2 artists. We are interested in choosing at most one person of each profession to create of representative group of five people. There are...

T1	T2	T3			3 ways to choose one teacher.
L1	L2	L3	L4		4 ways to choose one lawyer.
D1	D2	D3			3 ways to choose one doctor.
E1	E2	E3	E4	E5	5 ways to choose one engineer.
A1	A2				2 ways to choose one artist.

There is also one way to choose no teacher, one way to choose no lawyer, etc. Therefore, a generating function that represents this scenario is

$$p(x) = (1 + 3x)(1 + 4x)(1 + 3x)(1 + 5x)(1 + 2x). \qquad (6.11)$$

(a) Expand the polynomial 6.11 by multiplying.

(b) You decide that four representatives is enough for the survey. How many ways can you choose four people that consist of at most one from each profession? Explain your answer.

(c) Rewrite the polynomial so that it traces the choice of teachers.

(d) How many groups of size four have a teacher represented? Explain your answer.

6.4 Polynomials and Codes

In today's world technology plays a central role. What might not be evident is the fact that wherever computers or computer chips are used codes are required to send them instructions. The language of technology is codes. A great deal of money and time are spent on developing codes that will protect the consumer as well as quickly process the information that is needed. Codes are used in everything from credit cards, account numbers, CD players, laser scanners, remote controls, etc. Just about any modern convenience you can name involves some type of code. This handout explores one of those simple codes.

6.4.1 An Example of Modulo Arithmetic

The mathematics we will discuss in this section is known variously as **clock arithmetic**, **remainder arithmetic** or **modulo arithmetic**. To see how it works we will begin with an example.

Example 141 *(Modulo 5): Consider the set of nonnegative integers* $\mathbf{Z}^+ = \{0, 1, 2, 3, ...\}$*. We want to divide each integer by* $n = 5$ *and record the remainder. Since there can only be 5 remainders, either* $0, 1, 2, 3$ *or* 4*, then we can collect the integers together into sets of equal remainders.*

$$\{0, 5, 10, 15, 20, ...\} \text{ - all numbers of the form } 5k + 0.$$
$$\{1, 6, 11, 16, 21, ...\} \text{ - all numbers of the form } 5k + 1.$$
$$\{2, 7, 12, 17, 22, ...\} \text{ - all numbers of the form } 5k + 2.$$
$$\{3, 8, 13, 18, 23, ...\} \text{ - all numbers of the form } 5k + 3.$$
$$\{4, 9, 14, 19, 24, ...\} \text{ - all numbers of the form } 5k + 4.$$

Since it is the remainder that makes the difference with each set we will represent each set above as $\overline{0}, \overline{1}, \overline{2}, \overline{3}$ *and* $\overline{4}$*, respectively. Thus, we have a new set with but five elements* $\{\overline{0}, \overline{1}, \overline{2}, \overline{3}, \overline{4}\}$*.*

6.4.2 Modulo Arithmetic Operations

We can now define a type of addition and multiplication on the set $\{\overline{0}, \overline{1}, \overline{2}, \overline{3}, \overline{4}\}$.

Addition:

Note that $\overline{2} + \overline{4}$ means a number whose remainder is 2 when divided by 5 plus a number whose remainder is 4 when divided by 5. That is,

$$(5k + 2) + (5t + 4) = 5k + 5t + 6 = 5(k + t + 1) + 1 = \overline{1}$$

Another example is $\overline{3} + \overline{1}$ which means a number whose remainder is 3 when divided by 5 plus a number whose remainder is 1 when divided by 5. That is,

$$(5k + 3) + (5t + 1) = 5(k + t) + 4 = \overline{4}$$

Since there are a finite number of additions we can create an "addition" table of all possible additions.

Exercise 142 *Fill in the table below with all the other additions.*

Modulo 5 +	$\overline{0}$	$\overline{1}$	$\overline{2}$	$\overline{3}$	$\overline{4}$
$\overline{0}$					
$\overline{1}$				$\overline{4}$	
$\overline{2}$					$\overline{1}$
$\overline{3}$		$\overline{4}$			
$\overline{4}$			$\overline{1}$		

Multiplication

Multiplication is done in a similar fashion. For example, $\overline{2} \times \overline{4}$ means a number whose remainder is 2 when divided by 5 times a number whose remainder is 4 when divided by 5. That is,

$$(5k+2)(5t+4) = 25kt + 20k + 10t + 8 = (25kt + 20k + 10t + 5) + 3$$
$$= 5(5kt + 4k + 2t + 1) + 3 = \overline{3}$$

Exercise 143 *Fill in the modulo 5 table for multiplication.*

Modulo 5 ×	$\overline{0}$	$\overline{1}$	$\overline{2}$	$\overline{3}$	$\overline{4}$
$\overline{0}$					
$\overline{1}$					
$\overline{2}$					$\overline{3}$
$\overline{3}$					
$\overline{4}$			$\overline{3}$		

Exercise 144 *Create the addition and multiplication tables for modulos 6, 7 and some other numbers of your choice.*

Modulo 6 +	$\overline{0}$	$\overline{1}$	$\overline{2}$	$\overline{3}$	$\overline{4}$	$\overline{5}$
$\overline{0}$						
$\overline{1}$						
$\overline{2}$						
$\overline{3}$						
$\overline{4}$						
$\overline{5}$						

Modulo 6 ×	0	1	2	3	4	5
0						
1						
2						
3						
4						
5						

Modulo 7 +	0	1	2	3	4	5	6
0							
1							
2							
3							
4							
5							
6							

Modulo 7 ×	0	1	2	3	4	5	6
0							
1							
2							
3							
4							
5							
6							

Exercise 145 *What do you notice about the multiplication table for modulo 6 that is different from those for modulo 5 and 7?*

6.4.3   Applications

There are all sorts of interesting things that we can do with this new arithmetic. For example, the polynomial $x^2 + x + 1$ is not factorable in regular integer arithmetic but it is factorable modulo 7 arithmetic:

$$x^2 + x + 1 = (x + 5)(x + 3)$$

Note that the "numbers" 1, 5 and 3 in this factoring problem are really the "remainders" of some numbers. That is why it is possible to factor this quadratic in modulo 7. The process is actually quite easy. Since there are only 7 "numbers" modulo 7, we simply substitute each of the numbers $\{0, 1, 2, 3, 4, 5, 6\}$ into the quadratic until we find at most two of the numbers that produce a "zero" answer.

$x = 0$	$0^2 + 0 + 1 = 1$	$1 (\bmod 7)$
$x = 1$	$1^2 + 1 + 1 = 3$	$3 (\bmod 7)$
$x = 2$	$2^2 + 2 + 1 = 7$	$0 (\bmod 7)*$
$x = 3$	$3^2 + 3 + 1 = 13$	$6 (\bmod 7)$
$x = 4$	$4^2 + 4 + 1 = 21$	$0 (\bmod 7)*$
$x = 5$	$5^2 + 5 + 1 = 31$	$3 (\bmod 7)$
$x = 6$	$6^2 + 6 + 1 = 43$	$1 (\bmod 7)$

Therefore, $x^2 + x + 1 = (x - 2)(x - 4)$. However, it is customary to rewrite the factored form without negative numbers since we wish to remain within the set $\{0, 1, 2, 3, 4, 5, 6\}$. So we "add" 7 to each until the result is positive:

$$x^2 + x + 1 = (x - 2)(x - 4) = (x + 5)(x + 3).$$

A common application of modulo arithmetic to codes is seen in checkout laser scanners and CD players. Each item in stores have a "bar code" placed on its packaging. The number usually identifies the particulars of the item (size or quantity, manufacturing location, model type, flavor, etc.) When purchasing an item the cashier passes the bar code across the laser so it can "read" the identifying bar code into the system. In a sense, this is how the cashier "communicates" with the cash register.

Any communication system requires two things: both parties (the transmitter and the receiver) understand the "language being used, and a method by which errors can be identified and corrected. When the system consists of two people in the same room both able to understand the same language, then the communications channel is nothing more than speaking to each other. Since message transmission is very easy and quick error-correction is nothing more than retransmission. That is, if a plane flies overhead as one party is speaking then the listening party need only ask the speaker to say their message again. However, in the case of satellite communications asking a far off satellite to repeat a message is not very efficient or practical. In the case of the laser scanner at a check out counter it is easy to retransmit the data by simply re-scanning the bar code. Therefore, we don't need an elaborate error-correcting code. We only need to know if the item was scanned correctly or not. We will demonstrate one possibility for encoding identifications of items by way of example.

Example 146 *Casting out elevens. Suppose an item has identification number 7598307.*

1. *We will write this number in expanded form*

$$7 \times 10^0 + 5 \times 10^5 + 9 \times 10^4 + 8 \times 10^3 + 3 \times 10^2 + 0 \times 10^1 + 7 \times 10^0$$

2. *If we replace the "10" above with an x then we have a polynomial of the form*

$$\begin{aligned} p(x) &= 7 \times x^0 + 5 \times x^5 + 9 \times x^4 + 8 \times x^3 + 3 \times x^2 + 0 \times x^1 + 7 \times x^0 \\ &= 7x^0 + 5x^5 + 9x^4 + 8x^3 + 3x^2 + 7 \end{aligned}$$

3. *Clearly,* $p(10) = 7598307$.

4. *Note that* -1 *is the same "number" as* 10 *when we consider it as modulo* 11. *So,* $p(10) = p(-1)(\mod 11)$.

5. *This also means that* $p(-1)$ *alternately adds and subtracts the digits of the identification number*

$$7 - 5 + 9 - 8 + 3 - 0 + 7.$$

Can you explain why?

6. *The resulting number* $7 - 5 + 9 - 8 + 3 - 0 + 7 = 13$ *is also written modulo* 11. *That is*

$$7 - 5 + 9 - 8 + 3 - 0 + 7 = 2(\mod 11).$$

7. *This answer is added to the end of the identification number to create the bar code for the product*

$$75983072.$$

8. When the bar code is scanned by the laser it performs this operation of alternately adding and subtracting the bar code number, including the '2' at the end. If the answer it gets is '0' (that is, a multiple of 11) then it knows the item was scanned correctly. It then stores that number in the cash register and flashes a green light to let the customer and cashier know it has been done. If the answer it gets is not '0' then it knows the item was scanned incorrectly. It then flashes a red light to let the cashier know to retransmit, that is to scan the item again.

6.5 Summary

Polynomial models have been shown to be quite useful. Earlier chapters discussed some of the applications of linear and quadratic models. Both linear and quadratic models are special cases of polynomial models. In this chapter applications of higher order polynomials were discussed. These higher order polynomials can be used to aid in decision making or in enumerating possible choices in several situations.

The examples in this chapter concentrated on discrete-type examples of applications. However, there are many other ways in which polynomials can be applied to many other fields. It is left to the reader to explore some of these additional examples on their own.

Chapter 7

Exponential Models

7.1   Introduction

A recent report by the National Center for Health Statistics (NCHS) details an increase in the number of multiple births in the United States since 1970. The NCHS report implies that the increase in the number of multiple births, defined as babies born in sets of three or more, is due to the use of fertility drugs. The chart 7.1 below shows the increase in the number of multiple births since 1970.

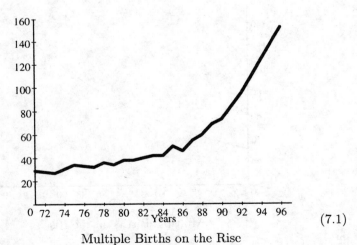

(7.1)

Multiple Births on the Rise

SOURCE : National Center for Health Statistics

The curve that appears to take shape in the graph 7.1 takes on a definite, and common, shape. The curve is known as an **exponential curve**. Exponential models are very useful when studying the dynamics of population growth and decay. A population is defined as a set of objects that is subject to changes

203

in size, such as colonies of bacteria, insects and animals as well as financial accounts or levels of chemical elements in the atmosphere. Because of the wide variety of such populations the exponential model is very important to modern mathematical applications.

Definition 147 *An **exponential function model** is one of the form*

$$f(t) = Pa^{kt}$$

*where P is the population size at time $t = 0$, a is the base of the function model and k, if positive, is the **intrinsic growth rate** of the population, and, if negative is the **intrinsic decay rate** of the population. If the base of the model is $a = e = 2.71828...$, then*

$$f(t) = Pe^{kt}$$

*is called the **natural exponential function model**.*

If this model is applied to the data used in chart 7.1 the resulting graphical model is

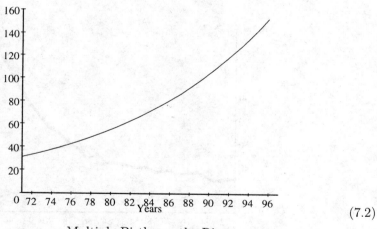

Multiple Births on the Rise

(7.2)

The analytical model that accompanies this curve is

$$f(t) = .336e^{.0637t}.$$

Although higher mathematics is needed to explain why the exponential model occurs within population dynamics it is still possible to study its use without going into any of those details. This chapter will take a look at some of the applications of the exponential model.

7.2 Financial Investment Models

7.2.1 Start Investing Early: Compound Interest

Most people believe that a large sum of money is needed in order to start an investment. Actually with the opportunities in today's markets nothing could be further from the truth. In fact, not only are there many opportunities to start an investment with as little as a few hundred dollars but many of these opportunities can now be found directly on the internet. Most firms advertise information about their investment policies on the internet along with instructions on how to invest. Even in the case of students there are investment opportunities. The sooner an investor starts the better the payoff in the end.

To show this let's review the idea of **compound interest**. When money is invested in a savings account, a money market account, mutual fund, bond market, stocks, etc. in a sense the investor is "loaning" the investment company their money so that the investment company can "re-loan" it to others interested in starting businesses, buying real estate, etc. Of course, the investment company agrees to pay the original investor a "fee" for the use of your money. This fee is the interest they pay you in regular intervals. The number of times your money is **compounded** is the number of times interest is added to your account. Naturally, the "bank" makes its money by charging higher interest rates to businesses and home buyers to whom they lend the money. This is the "fee" they pay for the use of the "bank's" money in buying their houses, etc.

Different types of accounts offer different amounts of interest. Interest on an account is determined by several factors: the accessibility of the account (how long you agree to keep the money in the account without withdrawing any of it); the "risk" involved (stocks have a higher risk of failure than government securities); etc. In the simplest of terms, the investment house provides an **annual rate of interest**, represented here by the letter r, the number of times the account is **compounded**, represented here by the letter k, the number of years you intend to keep the money in the account, represented here by the letter t, the initial amount you intend to invest (called the **principle amount**), represented here by the letter P, and the current amount in the account, represented here by the letter A. The interest that you are paid during each compounding is $\frac{r}{k}$, that is, if there are k compoundings each year, you earn $1/k^{th}$ of the yearly interest added to your account each compounding. From this information a general formula can be developed.

Example 148 *The interest earned on an account is calculated at the end of each period:*

1. *__End of first period of investment:__*

$$A = [current\ amt.] + \{interest\} = [P] + \left\{ \frac{r}{k} \bullet P \right\} = P(1 + \frac{r}{k})$$

2. **End of second period of investment:**

$$
\begin{aligned}
A &= [current\ amt.] + \{interest\} \\
&= \left[P(1+\frac{r}{k})\right] + \left\{\frac{r}{k} \bullet P(1+\frac{r}{k})\right\} \\
&= P(1+\frac{r}{k})[1+\frac{r}{k}] = P(1+\frac{r}{k})^2
\end{aligned}
$$

3. **End of third period of investment:**

$$
\begin{aligned}
A &= [current\ amt.] + \{interest\} \\
&= \left[P(1+\frac{r}{k})^2\right] + \left\{\frac{r}{k} \bullet P(1+\frac{r}{k})^2\right\} \\
&= P(1+\frac{r}{k})^2[1+\frac{r}{k}] = P(1+\frac{r}{k})^3
\end{aligned}
$$

From this pattern a general formula can be deduced for the kth period of investment.

4. **End of kth period of investment:**

$$
A = P(1+\frac{r}{k})^{kt}
$$

This gives the following

Definition 149 *The amount of money A in an account earned on a principle P with annual interest rate r compounded k times per year over t years is given by*

$$
A = P(1+\frac{r}{k})^{kt}. \tag{7.3}
$$

Here is a simple example:

Example 150 *A **certificate of deposit (CD)** is an account with an investment institution in which the investor agrees not to withdraw any money for a specified period of time (6 months, 12 months, 18 months, 24 months, 36 months, etc.). Early withdrawal results in penalties, however, the reward is a higher interest rate: the longer the specified time period the greater the interest rate. Most CD's require a minimum investment of $500.*

A local credit union offers 36 month CD's at 6.5% annual interest ($r = .065$) compounded monthly (12 times per year).

If an amount of $500, the minimum required, is invested, then in three years the account will be worth

$$
A = 500(1+\frac{.065}{12})^{12\cdot 3} = \$607.34
$$

Note that the Amount Formula above has several variables in it. If it is to be thought of as a function then we will have to concentrate on two of the variables and their relationship to each other while the other values remain constant.

Exercise 151 *If $P = \$1000, r = 8.4$ and $k = 12$, we wish to compare how the total amount A (co-domain) is affected by the years it is invested t (domain) then the Amount Formula is written as $A(t) = 1000(1.007)^{12t}$. Fill in the table below with the appropriate values.*

Years (t)	3	6	9	12	15	18
Account Amount (A(t))						

Exercise 152 *If $t = 5, r = 8.4$ and $k = 4$ (compounded quarterly), we wish to compare how the total amount A (co-domain) is affected by the principle initially invested P (domain) then the Amount Formula is written as $A(P) = P(1.021)^{60}$. Fill in the table below with the appropriate values.*

Principle (P)	500	1000	1500	2000	2500	3000
Account Amount (A(P))						

7.2.2 Start Investing Early: Periodic Deposits

Of course, a wise investor will want to consider several options. It is not realistic with today's investors to initially invest a large sum of money and then ignore the account for an extended period of time without adding to the account from time to time. Usually, when an account is started it is started with a smaller principle, say $500 or less, and then the investor makes arrangements to have money automatically transferred from a checking/savings account to the investment account each month. This is an excellent way to build up a larger account over a reasonable span of time. In fact, many people use this approach to anticipate the purchase of large items such as houses, cars, or tuition. It behooves the college student to begin such investments as freshmen, using smaller amounts of money earned through part time "college" jobs to anticipate future tuition needs.

Example 153 *Suppose that a freshman places $500 into an account that invests in government securities. Ginnie Maes are a form of government securities that invest in housing mortgages. Suppose that Ginnie Maes are currently paying an average annual rate of 8.5% compounded monthly (many investment houses have a government securities branch with current rates). You also decide to add an additional $25 each month to the account. This can be done by either the investor remembering to make the deposit each month or to give the investment house permission to deduct $25 each month automatically from the investor's checking account. Thus, after each month a new amount is calculated for the account. The first few months are listed here:*

1. **End of first month of investment:**

$$A_{one\ month} = A(1) = 500(1 + \frac{.085}{12}) = \$503.54$$

2. **End of second month of investment:** *Note that a new deposit of $25.00 is made during the second month. This $25.00 deposit will earn interest over the second month while the original $500.00 earns interest over the first **and** second months. The total amount of the investment after two months is represented by the formula*

$$A_{two\ months} = A(2) = 500(1 + \frac{.085}{12})^2 + 25(1 + \frac{.085}{12}) = \$532.29$$

In the space below explain the purpose of the values 500, 25, .085. 12 and the reason for the exponents in each term.

3. **End of third month of investment:** *Note that a new deposit of $25.00 is made during the third month. This $25.00 deposit will earn interest over the third month while the original $500.00 earns interest over the first, second **and** third months, and the $25.00 investment from the second month is earning interest from the second and third months. The total amount of the investment after three months is represented by the formula*

$$A_{two\ months} = A(2) = 500(1 + \frac{.085}{12})^3 + 25(1 + \frac{.085}{12})^2 + 25(1 + \frac{.085}{12}) = \$561.23$$

In the space below explain the purpose of the values 500, 25, .085. 12 and the reason for the exponents in each term.

4. ***End of fourth month of investment:*** *Use the space below to calculate the amount in the investment account after four months. Explain your results.*

5. ***End of three years (36 months) of investment:*** *If $25.00 is continued to be deposited each month into this account, then after three years (36 months) the account will have approximately*

$$A(36) = 500(1 + \frac{.085}{12})^{36} + 25(1 + \frac{.085}{12})^{35} + 25(1 + \frac{.085}{12})^{34} \qquad (7.4)$$
$$+ 25(1 + \frac{.085}{12})^{33} + \cdots + 25(1 + \frac{.085}{12})^{0}$$
$$= 500(1.007)^{36} + 25(1.007)^{35} + 25(1.007)^{34} + 25(1.007)^{33} + \cdots + 25(1.007)^{0}$$
$$= 500(1.007)^{36} + \sum_{k=0}^{35} 25(1.007)^{k} = \$1662.23$$

In three years with the modest start of $500 and $25 per month the account is worth $1662.23. Over that time the investor has invested $500 + (\$25)(35) = $1275 of their own money, which means that the account has earned approximately $388 in interest.

Exercise 154 *Take a moment in class to discuss how formula 7.4 was obtained.*

An investor generally does not ask how much money will be in an account after so many years. Rather the investor wants to know how much should be invested now and how much of a monthly deposit should be made so that after so many years there will be a specific amount in the account. That is, the investor thinks of the principle as a function of the end amount, $P(A)$, or

$$P(A) = \frac{A}{(1 + \frac{r}{k})^{kt}}. \tag{7.5}$$

Exercise 155 *An investor wants to have $5000 in an account after five years. If the investor decides go with the Ginnie Mae Securities (8.5% compounded monthly) discussed above then, without a continual monthly deposit added to the account over the five years, the investor would need*

$$P(5000) = \frac{5000}{(1.007)^{60}} = \$3290$$

or approximately $3300 as an initial investment. Fill in the table

Account Amount (A)	5000	10000	15000	20000	25000	30000
Principle (P(A))						

Take a moment to consider the following questions.

1. *Obviously we all do not have $3300 handy to invest today. Take a moment in class to determine how much of an initial investment it would take to have $5000 in five years if monthly deposits of $25.00 are made. (Note: trial and error is OK to use here)*

2. *Determine what the monthly deposits would have to be if the investor wants $5000 in five years and the investor initially invests $500. (Note: trial and error is oK to use here)*

7.2.3 Carbon Dating

Another common application of exponential growth and decay is **carbon dating**. Carbon dating is a technique for discovering the age of ancient objects (such as bones, clothing, furniture, etc.) by measuring the amount of Carbon 14 that the item contains. While plants and animals are alive the amount of Carbon 14 content within their system remains constant. However, when they die it decreases due to radioactive decay.

Scientists have spent many years studying the rate of Carbon 14 decay. As a result of this data they have determined that the amount, $P(t)$, of Carbon 14 in an object t years after it dies is given by

$$P(t) = (15.3)e^{-0.12104t} \approx (15.3) \times (0.886)^t.$$

The quantity $P(t)$ measures the rate of Carbon 14 atom disintegrations and this is measured in "counts per minute per gram of carbon (cpm)."

Exercise 156 *An archeologist has just been presented with two wooded objects. One is an artifact that is 4000 years old dating from ancient Egypt and the other is a new replica made from a fresh tree.*

1. *How much Carbon 14 does each sample contain? Explain. (Answer in cpm's)*

2. *How long does it take for the amount of Carbon 14 in each sample to be halved? That is, what is the half life of Carbon 14 in each object? These two answers should be the same. Explain why that is.*

3. *The first two questions are obviously backwards. Archeologists are gener-*
    ally interested in finding out how hold old an artifact is which means they
    already know how much Carbon 14 is present. Charcoal from the famous
    Lascaux Cave in France gave a count of 2.34 cpm. Estimate the date of
    formation of the charcoal and give a date to the paintings found in the
    cave.

4. *In the mid-1980's the Pope gave permission to test the age of the shroud of*
    Turin. Scientists discovered that the amount of Carbon 14 in the shroud
    was approximately 1.354157×10^{-41} cpm. How likely is it that this was the
    burial shroud of Jesus as church officials claim?

5. *Bones A and B are x and y thousands of years old, respectively. Bone A*
    contains three times as much Carbon 14 as bone B. Write a few sentences
    on what can be said about x and y?

7.2.4 Problems

1. When reports use mathematical and statistical results it behooves us to read those articles carefully to be certain the mathematics and statistics are saying what is implied. The NCHS report on multiple births found in section 7.1 also included the following statistics:

   - A record 6,000 babies were born in sets of three, four or more in 1996, a one year leap of 19 percent.

   - A continued drop in the number of women smoking while pregnant, although the number of pregnant teenagers who smoked increased. Overall, 13.6 percent of pregnant women smoked, down steadily from 20 percent in 1989.

   - An increase, for the seventh year running, in the number of women receiving early prenatal care. Improvements were tallied for women in all racial and ethnic groups, with more than eight in 10 women seeing a doctor during the first trimester. Teenagers were least likely to get this important early care.

   - A small drop in the percent of all out-of-wedlock births, down 1 percent. About, 1.2 million children were born to unmarried mothers, or about one-third of all births.

   - Births to unmarried black women have dropped 18 percent since peaking in 1989. A total of 7.4 percent of unmarried black women of childbearing age gave birth in 1996. The rate for Hispanic women was 9.3 percent, and for white women, 2.8 percent.

   - In 1996, there were 100,750 babies born as twins, 5,298 as triplets, 560 as quadruplets, and 81 as quintuplets or more.

   - Just 6 percent of single births result in low-birth-weight babies. For twins, it's 53 percent, and for triplets 93 percent. And just 8 percent of single births are premature compared with 53 percent for twins and 92 percent for triplets.

   - In 1995, 3 percent of babies died in their first month overall, but for multiple births, the figure was 16 percent.
     Use this information to answer each of the following:

   (a) Explain why it is necessary to record the number of multiple births as the number of higher-order multiples per 100,000 in the chart 7.1 above instead of simply as the total number of higher-order multiples for each year.

   (b) The data states that from 1995 to 1996 there was a 19 % increase in the rate of higher-order multiple births. Use chart 7.1 to check if this statistic is correct.

   (c) The fourth bullet point states that about 1.2 million children were born to unmarried mothers, or about one-third of all births in 1996.

Use this information to determine the total number of children born in 1996. Does this agree with the statement made prior to the bullet items?

(d) Bullet item two states that in 1996 13.6% of all pregnant women smoked during pregnancy. Use this information to determine the total number of pregnant women in 1996.

(e) Estimate the number of single birth low-birth-weight babies in 1996.

(f) Estimate the number of low-birth-weight babies born as a twin in 1996.

(g) Estimate the number of low-birth-weight babies born as a triplet in 1996.

(h) Estimate the number of multiple birth babies that died in their first month in 1995.

2. As a student your finances may be limited, however, most of us maintain part-time jobs to help pay for tuition and other expenses. Think of investing as one of the bills that get paid each month. With this in mind,

(a) Do a search on the internet for an investment company that specializes in small investments or student investments (many of these involve government securities of one sort or another). Identify an account and its investment return (annual interest rate).

(b) Determine an amount of money that you feel, as a student, you could comfortably invest to open the account ($100, $250, etc.) as well as an additional "payment" amount that you continually add on a monthly basis.

(c) If the interest is compounded monthly, calculate what the account would be worth in

   i. three years.
   ii. five years.
   iii. ten years.

3. We all think about purchasing big ticket items: a new car, a house, taking a vacation, getting a head start on investing or retirement. Take a moment to think about where you'd like to be five years from now as far as a "cash reserve" is concerned.

(a) Determine what for you would be a reasonable cash reserve in five years.

(b) Using the account information from problem 1 above determine an initial investment that,if invested in the above account, will produce this "reasonable" cash reserve in five years.

(c) If the answer in part b is unreasonable for you to invest initially, determine a monthly investment payment schedule that would produce the desired cash reserve yet be reasonable for you to maintain.

(d) Repeat parts a, b and c for

    i. ten years.

    ii. twenty years.

4. At the time of this writing, a regional credit union was offering a 5.00% rate compounded monthly on their 3-month certificate of deposit (CD) and a 5.75% rate compounded monthly on their 24-month CD. Naturally the rate for the 24-month CD is higher because you will be unable to withdraw the money from the account for two years.

(a) Determine the annual percentage yield, that is, the total percent of interest earned on the original principle at the end of the year.

(b) If you invested $500 in the 3-month CD, what would be the certificate's worth at the end of the three months?

(c) If you invested $500 in the 24-month CD, what would be the certificate's worth at the end of the three months?

(d) Obviously the advantage to the 3-month CD is quicker access to your money whereas the advantage to the 24-month CD is the greater interest rate. Devise an plan to invest your money in CD's that would give you the accessibility of a 3-month certificate with the interest rate of a 24-month certificate.

7.3 Summary

Exponential function models,

$$f(t) = Ae^{kt},\tag{7.6}$$

are some of the most applicable models in mathematics. One of the most common applications of exponential functions is to population dynamics, the study of the growth and/or decay of a population size. If the rate at which a population is changing in size is proportional to the population itself (say it doubles every 10 years for example) then this implies that the population growth is exponential. The "population" can be anything from the size of a spices to be studied to the amount of money in an account to the number of marbles in a bag. Population here is taken to mean the number of "objects" in a set to be studied.

The constant k in 7.6 is referred to as the intrinsic growth rate if $k > 0$ and the intrinsic decay rate if $k < 0$. This means that if $k > 0$ then the "population" is increasing and if $k < 0$ then the population is decreasing.

Note that $t = 0$ means when we begin our observations of a population or when we first have data for that population. It **does not** mean the beginning of time. For example, if we want to study the growth of the United States' population in the twentieth century then $t = 0$ is taken to be the year 1900. If we are only interested in the population in the second half of the century then $t = 0$ is taken to be 1950, etc.

Exponential models find many of their applications in fields that use Calculus and differential equations. The study of Differential Calculus includes the study of population changes which would naturally call for the use of exponential functions.

Additional Grids

This appendix contains some grids for in-class and homework use by the students and the instructor. Additional grids can be obtained by xeroxing those that appear here.

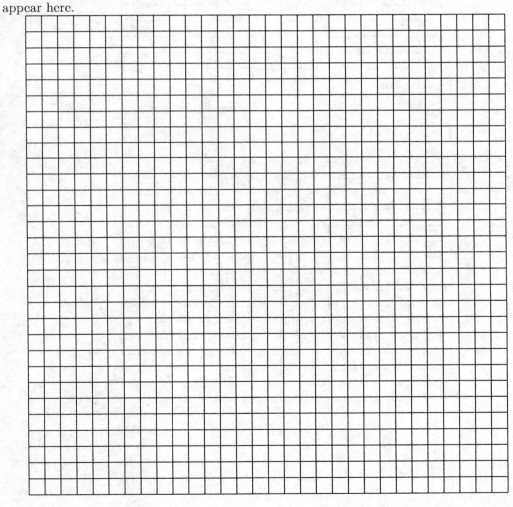

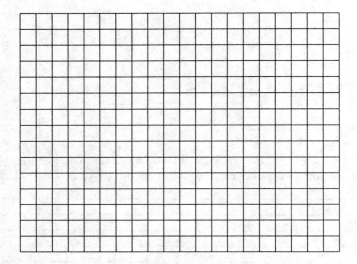

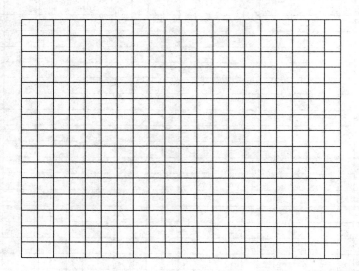

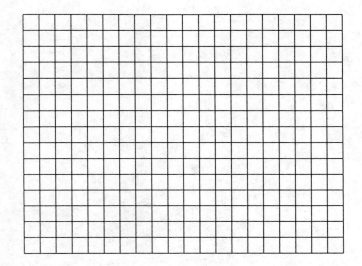

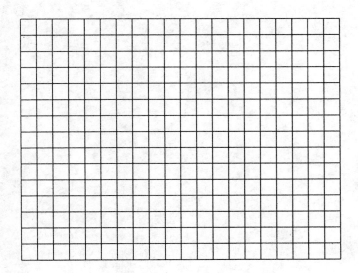

Additional Charts

This appendix contains some charts for in-class and homework use by the students and the instructor. Additional charts can be obtained by xeroxing those that appear here.

The following blank charts are for finite differences in Chapter 4.

| Term | Symbolic | Numerical | First | Difference | Second | Difference |
(n)	Expression	Value	Symbolic	Numerical	Symbolic	Numerical
			Blank	Blank	Blank	Blank
					Blank	Blank

| Term | Symbolic | Numerical | First | Difference | Second | Difference |
(n)	Expression	Value	Symbolic	Numerical	Symbolic	Numerical
			Blank	Blank	Blank	Blank
					Blank	Blank

Term (n)	Symbolic Expression	Numerical Value	First Symbolic	Difference Numerical	Second Symbolic	Difference Numerical
			Blank	Blank	Blank	Blank
					Blank	Blank

Term (n)	Symbolic Expression	Numerical Value	First Symbolic	Difference Numerical	Second Symbolic	Difference Numerical
			Blank	Blank	Blank	Blank
					Blank	Blank

Term (n)	Symbolic Expression	Numerical Value	First Symbolic	Difference Numerical	Second Symbolic	Difference Numerical
			Blank	Blank	Blank	Blank
					Blank	Blank

Index

arithmetic average, 157, 159

average, 155

binomial, 130

carbon dating, 211
cartography, 155
clock arithmetic, 195
co-domain, 23
coefficient, 179
completing the square, 145
compound interest, 205, 207
cosine, 99, 102

degree of a polynomial, 130
difference of cubes, 139
difference of squares, 139
domain, 23

examples, in-class exercises
    buried fuel tank, 10
    carbon dating, 211
    cellular phone bill, 48, 62
    cellular phone sales, 46
    certificate of deposit, 206
    chemical storage, 190
    compound interest, 207
    diversity of forest, 90
    efficency of metal detector, 133
    electric car test, 118
    financial comparison, 79
    function assignments, 20
    hand measurement, 12
    height of tree, 100
    identification codes, 199
    linear programming, 72
    lunar ice, 70
    measuring height, 97

    monitoring multiple births, 203
    monitoring stocks, 142
    oil well production, 43
    porosity, 93, 95
    quality control, 83
    safe driving (road rage), 108
    semester schedule, 188
    spread of a cold (disease), 4
    tagging, population size, 80
    tossing two fair dice, 183
    toxic spill, 161
    truck dispatcher, 39
    truck dispatcher model, 36
    truck rental fees, 51
    university waste, 168
    US population, 64
    wind velocity, 25, 115
exponent, 176
exponential model, 204
exponential properties, 176

factoring by grouping, 140
factoring techniques, 136
    difference of cubes, 139
    difference of squares, 139
    factoring by grouping, 140
    FOIL method, 136, 138
    sum of cubes, 140
finite differences, 115, 116, 119, 121
function
    co-domain, 23
    cosine, 99
    definition, 20, 23
    domain, 23
    greatest integer function, 63
    least integer function, 63
    range, 23

round-down function, 63
round-up function, 63
sine, 99
tangent, 100
function notation, 25

geometric average, 166
greatest integer function, 63

harmonic average, 166
horizontal axis, 155

intrinsic decay rate, 204, 216
intrinsic growth rate, 204, 216
inverse proportion, 108

least integer function, 63
like terms, 130, 178, 179
linear inequality model, 70
linear model, ii, 35, 62
linear programming, 72

mathematical model, ii
    analytical, 2, 9
    decisions, scheduling, 188
    definition, 1
    exponential, 203
    geometrical, 2, 3
    graphical, 2, 6
    linear, ii, 35, 62
    numerical, 2, 11
    polynomial, 175
    polynomial model, 130
    quadratic inequality, 142
    quadratic model, 115
    ratio and proportion, 79
midpoint formula, 159
modulo arithmetic, 195
monomial, 130
multiplying polynomials
    rectangular method, 133
    tree method, 132

natural exponential model, 204

ordered pair, 155

parabola, 118
percent, 85
percolation, 96
piecewise graph, 40
point-slope form of a line, 45

quadrant, 156
quadratic, 136
quadratic formula, 145
quadratic function model, 116

range, 23
rate of change, 36, 38, 48
ratio, 35
ratio and proportion, 79, 90, 93, 96
rectangular coordinate system, 155
regression, 65, 120, 168
remainder arithmetic, 195
round-down function, 63
round-up function, 63

sequence, 23
similar triangles, 98
    definition, 98
sine, 99, 102
slope
    definition, 45
slope of a line, 38, 43
    definition, 83
slope-intercept form of a line, 45
speed, 36, 38
step function model, 63, 65
sum of cubes, 140
system of equations, 66
    linear combination method, 68
    linear programming, 72
    substitution method, 68

tangent, 100, 102
translation of axes, 149
trigonometric ratios, 99
trinomial, 130

vertical axis, 155

weighted average, 158
    one-dimension, 158

three-dimension, 162
two-dimension, 160

y-intercept of a line, 50